THE ELEMENTS OF

MARIE CURIE

Also by Dava Sobel

The Glass Universe: How the Ladies of the Harvard Observatory Took the Measure of the Stars

And the Sun Stood Still: A Play

A More Perfect Heaven: How Copernicus Revolutionized the Cosmos

The Planets

Letters to Father: Suor Maria Celeste to Galileo, 1623–1633 (translated by Dava Sobel)

Galileo's Daughter: A Historical Memoir of Science, Faith, and Love

The Illustrated Longitude (with William J. H. Andrewes)

Longitude: The True Story of a Lone Genius Who Solved the Greatest Scientific Problem of His Time

THE ELEMENTS OF
MARIE CURIE

How the Glow of Radium Lit a Path for Women in Science

—

DAVA SOBEL

Atlantic Monthly Press
New York

FIRST EDITION

Published simultaneously in Canada
Printed in Canada

Book interior designed by Norman E. Tuttle at Alpha Design & Composition. This book is set in 12-pt. Bembo by Alpha Design & Composition of Pittsfield, NH.

First Grove Atlantic hardcover edition: October 2024

Library of Congress Cataloging-in-Publication data is available for this title.

ISBN 978-0-8021-6382-0
eISBN 978-0-8021-6383-7

Atlantic Monthly Press
an imprint of Grove Atlantic
154 West 14th Street
New York, NY 10011

Distributed by Publishers Group West

groveatlantic.com

24 25 26 27   10 9 8 7 6 5 4 3 2 1

To Harry Arthur Kobrin and Van Samuel Hix,
two future feminists

Contents

⁓

Preface

Formula for an Icon:
Marie Curie (1867–1934)

———

Even now, nearly a century after her death, Marie Curie remains the only female scientist whom most people can name.

She achieved her iconic status in spite of everything that stood in her way. Women were barred from university study in her native Warsaw, let alone from a career as a research scientist. Yet she succeeded in fully inhabiting that role. She inhabits it still.

The Nobel Prize in Physics, which she shared with her husband, Pierre, in 1903, and the Nobel Prize in Chemistry, awarded to her alone in 1911, surely helped immortalize her name. "Two-time Nobel Prize winner" continues to provide a ready shorthand explanation for Marie Curie's enduring fame. She was the first woman to receive a Nobel Prize, and the first person to win two of them. To date she remains the only Nobel laureate ever decorated in two separate fields of science.

Her twin prize medals, identically cast in solid gold, symbolize the gulf between the disparate ideals of womanhood and science. Each medal bears the bearded face of founder Alfred Nobel, the inventor of dynamite. On the reverse, two goddess figures in flowing robes mime a moment of discovery: "Science" raises her right hand to lift the veil from the visage of "Nature," who stands austere and bare-breasted, holding a cornucopia. The design consigns women to the realm of allegory, while the prizewinner's name engraved beneath the scene is typically that of a man in the mold of Alfred Nobel.

The 1903 Nobel Prize brought wealth as well as honor to the Curies, with a monetary reward of 70,000 gold francs. In return, Marie Curie helped burnish the luster of the Nobel Prize, which was then a novel phenomenon, first conferred in 1901. The storm of publicity that blew around her achievement spread both her name and the prize's name worldwide.

In 1906, after Pierre's untimely death in a senseless accident, the grieving Marie vowed to carry on their joint work. She stepped into her husband's place as director of the Curie laboratory and also took over his chair at the University of Paris, becoming that institution's first female professor. In her unique position, she could not help but attract numerous talented young women who wanted to work or study under her. They came from within France and also from abroad, as she had done when she left home to study at the Sorbonne. They included eager-to-be chemists and physicists from eastern Europe, Scandinavia, Russia, Great Britain, and Canada. As she nurtured her laboratory mentees, she also organized a small cooperative school for the sons and daughters of her friends, in which she taught a weekly physics class with hands-on activities.

By the time of her second Nobel, she was not only famous but infamous in the wake of a scandalous love affair. It took the Great War, which sent her to the front lines in a mobile X-ray unit of her own design, to restore her reputation as a heroine. In the 1920s, on two triumphal visits to America, she won new admirers by retaining her modest dress and direct manner even when speaking at large gatherings or meeting the president. Well before her death in 1934, she prepared her elder daughter to succeed her as laboratory director. Her younger daughter, who stayed with her at the sanitarium where she passed her final days, later told the illustrious, loving mother's life story in an acclaimed biography, called simply *Madame Curie*.

That account, and others that followed, made only passing mention of the forty-five aspiring female scientists who spent a formative
period in the Curie lab. Drawn, as Marie herself was, to the mystery
of radioactivity, and heedless of its dangers, they joined in discoveries, tested the power of radiation to treat disease, and explored the
unexpected world inside the atom as a source of limitless energy.
Several returned to their countries of origin to become the first
female professors there, or the first faculty to teach the new science
of radioactivity.

Given the social restrictions and expectations of their milieux,
some Curie protégées abandoned their careers when they married.
Others rejected marriage in order to pursue their research. A few
managed to combine the two. Those who befriended one another
at the lab later banded together in an international society devoted
to furthering educational and professional opportunities for women.
Long after their sojourns in Paris, they returned again and again to
a memory of some small moment in Mme. Curie's company—her
habit, say, of rubbing the tips of her radium-numbed fingers against
her thumb, or the way a smile would sometimes light her sad face
and render her suddenly beautiful.

Part One

School of Physics and Chemistry

42 Rue Lhomond, Paris

In the beginning was the Word, and the Word was . . . hydrogen.
—Diane Ackerman, *The Planets: A Cosmic Pastoral*

To extract uranium from pitchblende, there are at that time factories.
To extract radium from it, there is a woman in a hangar.
—Lydia Davis, "Marie Curie, So Honorable Woman,"
from *Samuel Johnson Is Indignant*

Chapter One

MANYA (Hydrogen)

THE WOMAN KNOWN to the world as Madame Curie entered the world on November 7, 1867, as Marya Salomea Sklodowska. The youngest of five children, she answered to Manya at home, as well as to the diminutive Manyusya, and sometimes Anciupecio. She had more pet names than any of her siblings because, in addition to being the baby of the family, she was also petite and precocious. She learned to read on her own by age four, and with such absorption that the older ones made a sport of scheming to distract her.

She owed her early interest in science to her father, Wladislaw Sklodowski, who taught mathematics and physics at a Warsaw high school for boys. He kept a barometer and other scientific apparatus in the family apartments, and communicated his enthusiasm for such things to his son and four daughters. He likewise revered language and literature. When he read aloud on Saturday evenings, he often chose a book in English, such as Charles Dickens's *David Copperfield*, translating the text into Polish on the fly. Selections in French or German required no such interpretation on his part for the sake of the children, who spoke several foreign tongues. He also inculcated in them a special reverence for the poets who celebrated the feats of Polish heroes.

The father's only fault, according to family lore, was his devotion to the lost cause of Polish nationhood. The country, once the largest in Europe, had been gradually overpowered by its aggressive neighbors, Austria, Prussia, and Russia, so that by 1795, all the land area long known as Poland had been partitioned and absorbed by the surrounding nation-states. Citizens who continued to consider themselves culturally Polish united to regain sovereignty, but the nationalist uprisings of 1830 and 1863 were brutally quashed, their leaders hanged, and thousands of followers exiled to Siberia. As a proud Pole teaching at a gymnasium in Russian-ruled Warsaw, Sklodowski provoked the anger of his superior, who fired him. Around that same time, unfortunately, the patriotic professor invested his life savings in his brother-in-law's business venture—a steam-powered mill in the countryside—which failed and ruined them both.

A deeper misfortune befell Manya's mother, Bronislawa Boguska Sklodowska, who had also taught school in Warsaw and served as headmistress of a private academy for girls. A few years after the birth of her fifth child, she developed the first signs of tuberculosis—a disease as deadly as it was common in those days. Though the routes of contagion were not well understood, she took every precaution she could think of, such as keeping separate dishes for herself. Fear of infecting her children forced her to stop hugging and kissing them. As the disease worsened, she sought rest and fresh air (the only available orthodox treatments) at mountain spas in Austria and southern France, accompanied by her eldest daughter, Zofia. For two years, beginning in 1874, Manya lamented the extended absences of these two beloved figures.

Meanwhile the cost of the mother's care drove the family into ever poorer lodgings. The desperate father took in more and more student boarders. These boys slept in the children's bedrooms, pushing Manya and her sisters into the living room, where still more boys came to study in the daytime. Under these overcrowded conditions, the two older Sklodowska girls, Zofia and Bronya, contracted typhus. Bronya recovered after several feverish weeks, but fourteen-year-old Zofia, who had for so long acted as nursemaid to her mother, died of the illness in January 1876. Their mother's death followed two years later, in May 1878, when Manya was ten.

"This catastrophe was the first great sorrow of my life and threw me into a profound depression," Mme. Curie disclosed decades later

in an autobiographical essay. "My mother had an exceptional personality. With all her intellectuality she had a big heart and a very high
sense of duty . . . Her influence over me was extraordinary, for in
me the natural love of the little girl for her mother was united with
a passionate admiration."

The remaining family members drew even closer together. Bronya
assumed many responsibilities of managing the household. The
bereaved husband preserved family traditions, including reunions
with relatives and friends, which "brought some joy" to all of them.

Manya had started school at the girls' academy on Freta Street—the
same one her mother had attended, later taught at, and ultimately
supervised. Manya had in fact been born in the Freta Street School,
near the center of Warsaw, in the apartment that came as a privilege
of her mother's position. The family vacated that home in 1868, while
she was still an infant. When she was old enough to return as a first-
grade student, she and her sister Helena, one year older, walked some
distance to reach the place from their Novolipki Street residence,
which lay west of the city center, on the border of the Jewish quarter.

At the beginning of third grade, in the fall of 1877, the two sisters
reported to a different school, closer to home. Because it was a private
school for girls, its bold administrator provided a real Polish education
in quiet defiance of the Russian authorities. Often the arrival of an
official inspector would send the students and teachers scrambling to
hide their Polish texts and feign engagement in one of the accepted
subjects, such as sewing or Russian history—in Russian. More than
once Manya, with her excellent command of the Russian language,
was singled out to answer an inspector's questions. "This was a great
trial to me, because of my timidity," she later wrote. "I wanted to
run away and hide."

The next school year, which began within months of her mother's
death, found her at Gymnasium Number Three, a government-run
girls' school where, as the grown Manya recalled, the teachers "treated
their pupils as enemies." Only from such a government-sanctioned
school, however, could she obtain a legal diploma. Now she walked
to school with her friend Kazia Przyborowska, whom she loved like

another sister, and after school she returned home with Kazia, whose mother coddled Manya and plied her with sweets.

"Do you know, Kazia," Manya wrote from a summer vacation at an uncle's farm, "in spite of everything, I like school. Perhaps you will make fun of me, but nevertheless I must tell you that I like it, and even that I love it. I can realize that now. Don't go imagining that I miss it! Oh no; not at all. But the idea that I am going back soon does not depress me, and the two years I have left to spend there don't seem as dreadful, as painful and long as they once did."

She was fifteen when she finished school on June 12, 1883, first in her class and winner of the gold medal—a distinction achieved earlier by her brother, Józef, and sister Bronya. Józef had gone directly from the boys' gymnasium to medical school at Warsaw University. Bronya would have loved to follow that same path, but it was open to men only, so she dreamed of someday studying medicine in Paris, where women were accepted at the Sorbonne. Helena possessed enough education as a gymnasium graduate to become a teacher like their mother, and enough musical talent to sing professionally if she chose to. Before Manya could form a clear picture of her own future, her father rewarded her scholastic success by sending her on a year's vacation with extended family, beginning with her mother's brothers in the southern countryside—the same relations who had embroiled him in the financial debacle. Now they treated Manya to the time of her life.

"I can't believe geometry or algebra ever existed," she wrote to Kazia. "I have completely forgotten them." The leisure activities she listed in her letters included reading "harmless and absurd little novels," walking in the woods, rolling hoops, playing childish games such as cross-tag and Goose, swinging "hard and high," swimming, fishing with torches for shrimp, horseback riding, and eating wild strawberries. She had the family dog, Lancet, with her in the country that summer, and reported his noisy misbehavior as part of her enjoyment. When the season turned, she went farther south to visit her father's brother in the foothills of the Carpathian Mountains, where she spent the next few months.

On festive winter nights Manya and her cousins, dressed as peasant girls, sped off through the forest in a sleigh, guided by young men on horseback. Other sleighs full of revelers met them at the first house they encountered, and a sleigh full of musicians, too. They danced the mazurka and the oberek, also the waltz, for hours. Then, instead of returning home, they rode on to the next house, and the next, in a custom called the *kulig*. Everywhere they stopped they found a feast prepared for them. Daybreak arrived before they finished their partying at all the hosts' homes.

Her year of freedom from care extended through the summer of 1884, when she and Helena were invited to the rural estate of one of their mother's former pupils. Up north in Kempa, the riverine scenery and hospitality topped even the amusements of the *kulig*. As she told Kazia, "Kempa is at the junction of the Narev and Biebrza rivers—which is to say that there is plenty of water for swimming and boating, which delights me. I am learning to row—I am getting on

Manya's portrait of Lancet in her diary

quite well—and the bathing is ideal. We do everything that comes into our heads, we sleep sometimes at night and sometimes by day, we dance, and we run to such follies that sometimes we deserve to be locked up in an asylum for the insane." On August 25, the morning after the St. Louis night ball, by Manya's own account, she discarded her new dance shoes because she had worn out their soles.

She returned to Warsaw, to her father's much smaller quarters. He still taught classes, but no longer took in boarders. Manya, nearing her seventeenth birthday, could now add her own wages to the family's support by giving private lessons in French, arithmetic, or geometry. It was not an easy way to make a living. "A person who knew of us through friends came to inquire about lessons," she noted. "Bronya told her half a ruble an hour, and the visitor ran away as though the house had caught fire."

To further their own education, she and Bronya attended secret sessions of the "Flying University." The limited course offerings of this itinerant institution depended on the expertise of its volunteer professors, who met with eight or ten students at a time in one or another private residence. Classroom locales shifted often to avoid detection by the police. While study topics such as anatomy and natural history could hardly be considered subversive, providing higher education to women was itself illicit under Russian rule.

In the spirit of the Flying University, Manya made regular visits to a dressmaking shop where she read aloud to the employees and built up a lending library for them of books in Polish. On her own she read fiction and philosophy, sketched flowers and animals (including Lancet) in a notebook, wrote verses, and figured out a plan that would enable her—and Bronya—to study in France.

She would get a job as a governess. Not only would she earn more money living with a family, but also her room and board would be covered, freeing the bulk of her wages—maybe as much as 400 rubles a year—to pay Bronya's way in Paris. In another five years, when Bronya became a doctor, then it would be Manya's turn to go to the Sorbonne, and Bronya's duty to support her.

Bronya, who was twenty, had saved just enough money to get to Paris and see herself through one year at the Faculty of Medicine. The plan fell into place in the autumn of 1885, just as Manya had envisioned, but by December her part of it had already fallen apart. Her placement with the B——s, a family of lawyers, made her feel like a prisoner. As she confided to her cousin Henrietta, "I shouldn't like my worst enemy to live in such a hell. In the end, my relations with Mme. B—— had become so icy that I could not endure it any longer and told her so. Since she was exactly as enthusiastic about me as I was about her, we understood each other marvelously well." Her presence in the B——s' home ended her innocence regarding human nature. "I learned that the characters described in novels really do exist, and that one must not enter into contact with people who have been demoralized by wealth."

She immediately found a new position. This one pulled her far away from home—to Szczuki, some fifty miles north of Warsaw—but placed her with a more tolerable family and paid an annual salary of 500 rubles. En route on January 1, 1886, she imagined she was headed to forested hills of the kind she had admired the previous summer. Instead she found herself surrounded by two hundred acres of agricultural fields, all devoted to growing beet root for sugar production. Indeed, the business of turning the harvested crops into sweetener took place in a dreary-looking factory adjacent to the home of the Zorawski family, her new employers. Past the factory a cluster of huts housed the peasants who worked on the land. The nearby river offered scant recreation, but served the factory and also received its waste. Still, after one month in residence, Manya let Henrietta know how much her situation had improved: "The Z.'s are excellent people. I have made friends with their eldest daughter, Bronka, which contributes to the pleasantness of my life. As for my pupil, Andzia, who will soon be ten, she is an obedient child, but very disorderly and spoiled. Still, one can't require perfection."

She worked seven hours a day—three with Bronka and four with Andzia. The antics of the younger children, a boy of three and

Manya and Bronya Sklodowska, 1886

a six-month-old baby girl, further brightened her spirits. She had yet to encounter the three older sons, who were away at boarding school and university in Warsaw.

Sometimes Mme. Zorawska begged Manya to help entertain guests by conversing with them or sitting in as a fourth at card games, and of course she complied. In her free time, of her own volition, she organized a class for the peasant children, teaching them for two hours a day to read and write in Polish, since they learned only Russian at school. Late at night and early in the morning, she pursued her own reading toward her ultimate goal of studying physics and mathematics at the Sorbonne. She named some of these books in a December letter to Henrietta: John Frederic Daniell's *Physics*, Herbert Spencer's *Sociology* in French, and Paul Bert's *Lessons on Anatomy and Physiology* in Russian. She was reading them all at the same time, she told Henrietta, because "the consecutive study of a single subject would wear out my poor little head which is already much overworked. When I feel myself quite unable to read with profit, I work problems of algebra or trigonometry, which allow no lapses of attention and get me back into the right road."

She let one disruption in her busy schedule go unmentioned in her letters to Henrietta: Sometime during her first year at Szczuki, she met and fell in love with the Zorawskis' oldest son, Kazimierz, the university student. Their romance led to a serious commitment, but when Kazimierz announced that he and Manya were engaged, his parents forbade him to marry the penniless governess. Kazimierz could not flout their wishes, and Manya, who could not afford to lose her income, swallowed her rage and shame and stuck to her work.

The episode darkened her view of the future. In March 1887, three months into her second year at Szczuki, when her brother was contemplating a medical office in the provinces, she begged him to hold out for something better in the big city. If he compromised, she said, then she would "suffer enormously, for now that I have lost the hope of ever becoming anybody, all my ambition has been transferred to Bronya and you. You two, at least, must direct your lives according to your gifts. These gifts which, without any doubt, do exist in our family must not disappear; they must come out through one of us. The more regret I have for myself the more hope I have for you."

By the following March she was even more despondent. "If only I didn't have to think of Bronya," she confessed to Józef, "I should present my resignation to the Z.'s this very instant and look for another post . . ." Despite the hopelessness and frustration she sometimes vented, she continued her own higher studies. "Think of it," she wrote Józef, "I am learning chemistry from a book. You can imagine how little I get out of that, but what can I do, as I have no place to make experiments or do practical work?"

At Easter in 1889, having fully discharged her duty to Andzia, she left Szczuki for another governess position, this time with the wealthy Fuchs family in Warsaw. She found living in their luxurious home pleasant enough, but was happy to leave it after one year to move in with her father once again and rely on giving private lessons for her income. When she re-enrolled at the Flying University, she found its student body had swelled from two hundred to one thousand women, and its classrooms had perforce relocated from scattered homes to various discreet institutions.

Through an older cousin on the Boguski side, she gained access for the first time to a real laboratory, located in central Warsaw at the Museum of Industry and Agriculture. In the evenings and on Sundays, by herself, she would go there to try out some of the experiments described in the treatises she was reading on chemistry and physics.

"I learned to my cost that progress in such matters is neither rapid nor easy," she later reflected. Even so, "I developed my taste for experimental research during these first trials."

———

MANYA'S ACCOMMODATING COUSIN, Józef Boguski, had studied chemistry in his youth at the University of St. Petersburg with Dmitri Mendeleev, creator of the periodic table of the elements. This remarkable chart summarized everything known about the building blocks of the material world. At a glance it showed which elements shared common properties, which ones were most likely to combine with which others, and in what proportions. Moreover, it gave scientific meaning to the age-old terms "atom" and "element."

"Atom" (*a-tom*) meant "un-cuttable" to ancient Greek philosophers pondering the smallest possible divisions of matter. By the nineteenth century "atom" indicated an invisible, indivisible particle, inconceivably tiny yet still retaining a given element's traits. "Element," through the long course of its history, had described a variety of essential entities, such as fire, air, earth, and water. Artisans in diverse early cultures smelted iron, alloyed copper with tin to make bronze, exploited the ornamental virtues of silver, and appreciated the utility of sulfur for cleaning, bleaching, and making medicines and matches. In the Middle Ages, alchemists endeavored to convert certain elements from a base form, such as lead, to the purity of gold. A list of thirty-three "simple substances" drawn up on the brink of the French Revolution added the gases hydrogen, oxygen, and nitrogen to the chemist's supply. Nineteenth-century discoveries of calcium, potassium, silicon, iodine, and a couple dozen others had increased the number of known elements to sixty-three by the time Mendeleev tried to impose some kind of order on the ever-growing assortment.

He chose atomic weight as his organizing principle. Although no scale could weigh an entity as minuscule as an atom, the preceding century of theory and experimentation had shown chemists a way around that problem: By weighing equal volumes of different gases, they found hydrogen to be the lightest of all, and assigned it an atomic weight of one. In the reaction between hydrogen and oxygen to yield water, the quantity of oxygen always outweighed that of hydrogen by a factor of eight, so the atomic weight of oxygen was set at eight—until chemists realized that water contained two atoms of hydrogen for every one of oxygen, and corrected oxygen's atomic weight to sixteen. Since both hydrogen and oxygen combined readily with many another element, these two helped establish the atomic weights of iron, sodium, magnesium, aluminum, and more.

Atomic weight proved to be an element's most enduring feature. It alone held fast while characteristic colors, textures, and odors typically disappeared when elements combined in chemical reactions. Sodium and chlorine, for example—one a soft, silvery-white metal, the other a poisonous gas—united to form crystals of ordinary table salt (sodium chloride), but their atomic weights remained unchanged.

— 70 —

но въ ней, мнѣ кажется, уже ясно выражается примѣнимость вы ставляемаго мною начала ко всей совокупности элементовъ, пай которыхъ извѣстенъ съ достовѣрностію. На этотъ разъ я и желалъ преимущественно найдти общую систему элементовъ. Вотъ этотъ опытъ:

```
                         Ti=50      Zr=90      ?=180.
                         V=51       Nb=94      Ta=182.
                         Cr=52      Mo=96      W=186.
                         Mn=55      Rh=104,4   Pt=197,4
                         Fe=56      Ru=104,4   Ir=198.
                       Ni=Co=59     Pl=106,6   Os=199.
H=1                      Cu=63,4    Ag=108     Hg=200.
     Be=9,4   Mg=24      Zn=65,2    Cd=112
     B=11     Al=27,4    ?=68       Ur=116     Au=197?
     C=12     Si=28      ?=70       Sn=118
     N=14     P=31       As=75      Sb=122     Bi=210
     O=16     S=32       Se=79,4    Te=128?
     F=19     Cl=35,5    Br=80      I=127
Li=7 Na=23    K=39       Rb=85,4    Cs=133     Tl=204
              Ca=40      Sr=87,6    Ba=137     Pb=207.
              ?=45       Ce=92
             ?Er=56      La=94
             ?Yt=60      Di=95
             ?In=75,6    Th=118?
```

а потому приходится въ разныхъ рядахъ имѣть различное измѣненіе разностей, чего нѣтъ въ главныхъ числахъ предлагаемой таблицы. Или же придется предпо- лагать при составленіи системы очень много недостающихъ членовъ. То и другое мало выгодно. Мнѣ кажется притомъ, наиболѣе естественнымъ составить кубическую систему (предлагаемая есть плоскостная), но и попытки для ея образо- ванія не повели къ надлежащимъ результатамъ. Слѣдующія двѣ попытки могутъ по- казать то разнообразіе сопоставленій, какое возможно при допущеніи основнаго начала, высказаннаго въ этой статьѣ.

```
Li   Na   K    Cu   Rb   Ag   Cs   —  Tl
7    23   39   63,4 85,4 108  133     204
Be   Mg   Ca   Zn   Sr   Cd   Ba   —  Pb
B    Al   —    —    Ur   —    —    —  Bi?
C    Si   Ti   —    Zr   Sn   —    —  —
N    P    V    As   Nb   Sb   —    Ta —
O    S    —    Se   —    Te   —    W  —
F    Cl   —    Br   —    J    —    —  —
19   35,5 58   80   190  127  160  190 220.
```

Mendeleev's first published periodic table
(in Russian)

When Mendeleev arrayed the known elements in order of ascending atomic weight, he was stunned to see certain chemical properties repeat at regular intervals. This periodic repetition of similarities convinced him he had perceived a law of nature.

Mendeleev published his periodic table in 1869 in a limited edition of two hundred copies, as well as in journal articles and in a textbook he wrote for his students at St. Petersburg. Since its debut, and thanks to Mendeleev's own frequent improvements over the following twenty years, his periodic table had become a laboratory fixture. The most current revised version available to Manya Sklodowska at the Museum of Industry and Agriculture included three elements only recently identified: gallium, scandium, and germanium, each named for its discoverer's homeland. All three occupied places on the periodic table that Mendeleev had intentionally left blank, as though awaiting their arrival. A few unusually large weight differences between successive elements had tipped him off to the possibility of latecomers. In making accommodation for them, he had given them provisional names—such as "eka-aluminum" for gallium—and approximated their atomic weights. Those predictions had since proven prophetic.

As Manya could see, empty spaces still punctuated the periodic table. At any moment, another new element might emerge to fill a vacant spot below tellurium, below barium, between thorium and uranium, or perhaps beyond uranium—the element deemed to be the heaviest.

News from Bronya in Paris came as fulfillment of the sisters' long-standing bargain. Bronya had married a fellow student, a Polish expatriate named Kazimierz Dluski, in the summer of 1890, and established a small medical practice with him. There was room in their apartment for Manya.

"Your invitation to Paris . . . has given her a fever," Wladislaw Sklodowski wrote to his elder daughter. "I feel the power with which she wills to approach that source of science, towards which she aspires so much."

On Bronya's advice, Manya shipped her mattress and other bulky belongings by freight, to save herself the trouble and expense of buying new ones in France. As a further economy, she chose the cheapest possible train fare—a mix of third- and fourth-class tickets that required her to bring her own folding seat for the section of the journey crossing Germany, plus enough food and drink to see her through three days' travel.

"So it was in November, 1891, at the age of twenty-four," she recalled three decades later, "that I was able to realize the dream that had been ever present in my mind for several years."

Chapter Two

MARIE (Iron)

———

NOTHING IN MANYA'S prior travels prepared her for the grandeur of Paris. Although Warsaw boasted its own royal palace, ornate churches, fine townhouses, and historic monuments, they all wore the drab aura of the Russian occupation. In Paris, she found the city's beauty heightened by the freedom of its citizens, who openly spoke their own language and discussed their ideals in public. Scientific research, which she had pursued in secret, here dominated numerous imposing spaces. The recently completed Louis Pasteur Institute was already drawing researchers from other countries and offering the world's first courses in microbiology. A grand Gallery of Zoology had opened near the Museum of Natural History inside the Botanical Gardens—which were not merely ornamental but maintained a centuries-old herbarium of medicinal plants. And the new Eiffel Tower, though derided by many as a desecration of the skyline, bore at its base the engraved names of seventy-two astronomers, chemists, physicists, naturalists, engineers, and mathematicians whom France proudly counted among her native sons.

The University of Paris, the Sorbonne, existed as an exalted city within the city. It boasted a student population of nine thousand, and its professors delivered their lectures dressed formally in white tie and cutaway jacket.

In the first week of November 1891, Manya enrolled in the *Faculté des sciences*—one of only twenty-three women, among nearly two thousand men, to do so. She made her name sound at least half French by signing her registration card as *Marie* Sklodowska, and quickly grew accustomed to being addressed as "Mademoiselle."

When evening came, however, and she returned home to her sister and brother-in-law, Paris disappeared. Stepping into the Dluskis' apartment on the rue d'Allemagne transported her back to Warsaw. Everything about the place, from its décor to the friends it attracted, evoked Poland. Only the doctors' patients were French. Bronya and her husband served the medical needs of a neighborhood called La Villette, near the slaughterhouses on the northeast outskirts of the city. Manya-Marie traveled from there to the Sorbonne and back via horse-drawn double-deck omnibuses, losing one full hour of study in each direction.

After a few months of this wearisome commuting, she decided her carfare could be better spent on renting a garret room in the Latin Quarter. In mid-March 1892 she wrote her brother, Józef, from her new lodgings at 3 rue Flatters. "It is a little room, very suitable, and nevertheless very cheap." She could walk to the chemistry laboratory in fifteen minutes, to the lecture hall in twenty.

"I am working a thousand times as hard as at the beginning of my stay," she told Józef. While living with the Dluskis, "my little brother-in-law had the habit of disturbing me endlessly. He absolutely could not endure having me do anything but engage in agreeable chatter with him when I was at home. I had to declare war on him on this subject." Peace between them had since been restored. "Naturally, without the Dluskis' help I should never have been able to arrange things like this."

In the thrill of her independence, Marie ignored the fact that she did not know how to cook. Meals had simply appeared before her in her childhood home and throughout her employment as a governess. Now, away from Bronya's table, she subsisted mainly on tea with buttered bread. She did not mind the limited diet, but soon her body rebelled in frequent dizzy spells. When she fainted in front of a Polish classmate, word reached the Dluskis, and Marie's "little brother-in-law" forcibly took her back to La Villette for a week of proper nutrition. Then, with exams looming, she returned to her garret.

"The room I lived in," she recalled in later years, "was in a garret, very cold in winter, for it was insufficiently heated by a small stove which often lacked coal. During a particularly rigorous winter, it was not unusual for the water to freeze in the basin in the night; to be able to sleep I was obliged to pile all my clothes on the bedcovers. In the same room I prepared my meals with the aid of an alcohol lamp and a few kitchen utensils. These meals were often reduced to bread with a cup of chocolate, eggs or fruit. I had no help in housekeeping and I myself carried the little coal I used up the six flights." Even so, the life held "a real charm" for her.

"All that I saw and learned that was new delighted me. It was like a new world opened to me, the world of science, which I was at last permitted to know in all liberty."

The forces that governed the universe—gravity, electricity, magnetism—permeated the lectures she attended, the experiments she attempted, and the treatises she pored over in the library or read in her room till all hours. She realized early on, however, that her years of earnest solo dedication had not adequately prepared her for university study in mathematics. Even her presumed fluency in French sometimes faltered. So she redoubled her efforts, pushed herself harder. At the end of two years, she placed first in her class at the July 1893 examinations and received her degree in physical sciences, the *licenciée ès sciences physiques*.

Her educational adventure might have ended there, but a 600-ruble Alexandrovitch Scholarship bought her another garret room, another year at the Sorbonne. "I hardly need say that I am delighted to be back in Paris," she wrote Józef in mid-September 1893 after a brief stay at home. "It was very hard for me to separate again from Father, but I could see that he was well, very lively, and that he could do without me—especially as you are living in Warsaw. And as for me, it is my whole life that is at stake."

She was studying "unceasingly," she said, in anticipation of the start of courses. This year she would work toward a mathematics degree. Her professors included physicist Gabriel Lippmann, who was at that time perfecting his theory of color photography, and famed mathematician Henri Poincaré.

Early in 1894 Professor Lippmann helped her secure a commission from the Society for the Encouragement of National Industry. She was to study the magnetic properties of dozens of varieties of steel—a matter of crucial commercial importance. Magnets had served as the essential components in navigational compasses for almost a millennium. More recently they had been put to wider use in electrical power technology. The electricity now coursing through the city of Paris enabled telegraph communication, powered streetcars, lit avenues at night, and sent elevators up and down the Eiffel Tower. New instruments were required to generate and measure the modern era's much stronger electric currents, and these tools relied on magnets at their core.

The steel investigation, Lippmann suggested, might prove an appropriate topic for a doctoral dissertation, should Mademoiselle choose to progress to the highest level of scholarly achievement. Clearly he judged her capable of such distinction. Meanwhile the work would pay her a small subvention. Mindful of her own inexperience, Marie accepted a friend's offer to introduce her to a potentially helpful physicist who was already well grounded in research on magnetism.

Marie's first impression of Pierre Curie—"a tall young man with auburn hair and large, limpid eyes" standing in the recess of a glass door—lodged indelibly in her memory. He was thirty-five at the time, eight years her senior, though she thought he looked younger than that. "I noticed the grave and gentle expression of his face, as well as a certain abandon in his stance, suggesting the dreamer absorbed in his reflections." He showed her "a simple cordiality." As their conversation flitted from science to social and humanitarian concerns that animated them both, they sensed a surprising kinship.

Neither one of them had come to this arranged meeting with the expectation of finding a life partner. She had been painfully spurned in love. He had long since sworn off intimacy. As he rationalized in some diary notes set down at age twenty-two, "A woman loves life for the living of it far more than we do. Women of genius are rare. Thus when we . . . give all our thoughts to some work which estranges us from those nearest us, it is with women that we must struggle. The mother wants the love of her child above all things, even if she should make an imbecile of him. The mistress also wishes to possess her lover, and would find it quite natural to sacrifice the rarest genius in the world for an hour of love."

Pierre Curie

They saw each other next at a meeting of the Physics Society, where Pierre was a regular fix-ture and a vocal participant in discussions. Soon afterward he sent her, care of Lippmann's laboratory where she worked on her project, a copy of his recent publication about the symmetry between electric and magnetic fields. He inscribed it, "To Mlle. Sklodowska, with the respect and friendship of the author, P. Curie."

"Some time later," she wrote in her autobiographical essay, "he visited me in my student room and we became good friends."

In French, the word for "magnet," *aimant*, also means "loving." Love makes an apt metaphor for the attraction of magnetic oppo-sites. Two bar magnets will cling together when their opposite poles are put in proximity but repel one another if like ends meet. As in the case of human allure, magnetic attraction may fade with time. In fact, although scientists of the late nineteenth century divided magnetic materials into "permanent" and "temporary" categories, they conceded that "permanent magnetism" was probably as elusive as everlasting love. Marie's assigned research, it was hoped, would expose those aspects of steel manufacture most favorable to the production of durable magnets.

Since ancient times, smelters had heated iron ore in charcoal fires to produce carbonized steel—a stronger, harder metal than iron for making weapons and tools. Modern French industrialists each followed a slightly different chemical formula, enriching their alloys with additional elements, such as chromium and manganese, to enhance one or another of steel's most desirable qualities. Manufacturers further differentiated their steel products by customizing their methods of heating and cooling the material, whether by annealing, quenching, or tempering. They guarded the details of these practices as trade secrets.

Marie received forty-seven samples of French steel for her study. A few of these were shaped in the form of rings, but most were small bars, twenty centimeters (about eight inches) long and one centimeter square in cross section. It fell to her to determine which ones could be most readily magnetized *and* prove most likely to retain their acquired magnetism through the years.

Pierre had conducted his own independent research on magnetism at the school where he worked as director of student laboratories. In one project, he demonstrated that materials such as iron and nickel lost their magnetic properties when heated, and he took care to note the temperature at which this change occurred for each of several materials.* Pierre was quite familiar with the apparatus required to induce magnetism in pieces of steel, to measure a given magnet's strength—in short, to subject magnets to almost any type of trial. He showed himself more than willing to share this knowledge with Marie.

Although it was uncharacteristic of either Marie or Pierre to allot much time to diversion, they somehow found opportunities to go places and do things together. One day he took her to Mi-Carême, the mid-Lenten carnival, where a jostling crowd carried Marie some distance away from him, so that minutes passed before they were reunited. The incident weighed on Pierre as a sign of how easily the

---

*These metal-specific degree readings are still known today as "Curie points."

bonds they were building might rupture. When he asked her to come meet his parents, at the home he still shared with them in the Parisian suburb of Sceaux, she fully grasped the import of the invitation.

She found Eugène and Sophie-Claire Curie to resemble her own father and mother in temperament, and she warmed to them as readily as she had to Pierre. Still, she could not commit herself to marrying a foreigner. As soon as she finished the academic term and received her *licenciée ès sciences mathématiques*—with the second highest rank in the exams this time—she hurried home to her own family.

That summer of 1894, separated by a thousand miles, she and Pierre corresponded regularly. "It would be a beautiful thing," he wrote to her in August, "to pass through life together, hypnotized by our dreams: your dream for your country; our dream for humanity; our dream for science. Of all these dreams, I believe the last, alone, is legitimate." Only in the realm of science, he explained, could they be certain of accomplishing more good than harm. "The territory here is more solid and obvious, and however small it is, it is truly in our possession."

A few days later he worried, "I do not know why I have got it into my head to keep you in France, to exile you from your country and family without having anything good to offer you in exchange for such a sacrifice." At the same time, he honestly believed that professional prospects for her were likely better in Paris than in Poland. If she would not marry him, might she agree to live with him in friendship? There were suitable rooms available "on the rue Mouffetard," he informed her, "with windows overlooking a garden. This apartment is divided into two independent parts."

In September she mailed him her photo, which pleased him "enormously." He showed her off to his older brother, Jacques, who had been his partner through years of productive research and the invention of several patented scientific instruments, such as analytical balances and meters to assess electric charge. Jacques, a mineralogy professor at Montpelier, thought Marie had "a very decided look, even stubborn."

When she did return to Paris, in October, she moved into a room at Bronya's new medical office, which stood empty outside of visiting hours. Pierre continued his active pursuit as best he could from Sceaux, about six miles south of Paris, where his mother now lay seriously ill, and was being cared for by his father, the local medical doctor.

"I'm not coming to see you tonight," Pierre apologized in a note canceling a midweek rendezvous. "My father has rounds to make and I will stay at Sceaux until tomorrow afternoon so that Maman won't be alone." In his uncomfortable insecurity, he added, "I sense that you must be having less and less regard for me while at the same time my affection for you grows each day."

For all his brilliance and originality, not to mention the numerous papers he had published in the physics journals, Pierre had never bothered to complete a doctoral degree. After his undergraduate studies at the Sorbonne, he had served as a laboratory assistant for one of the faculty members, then took the job he currently held—organizing and running the student laboratories at an industrial school. His pupils adored him, and the administration smiled on his independent research activities. Occasionally Pierre pursued a study of some natural phenomenon purely for the joy of intellectual exercise, with no attempt to report his results or claim credit for a discovery. He was so averse to self-promotion that he chose not to apply for a better position when one of the teachers at his school resigned, creating a vacancy.

"What an ugly necessity is this of seeking any position whatsoever," he complained in a letter to Marie. "I am not accustomed to this form of activity, demoralizing to the highest degree." When the head of school tried to gain him an official academic recognition, Pierre sent a no-thank-you note that read in part: "I am told that you intend to propose me again to the *préfet* for the decoration. I beg of you not to do so. If you win this distinction for me, you will oblige me to refuse it, for I am resolved not to accept any sort of decoration. I hope that you will want to help me avoid taking a step that will make me look somewhat ridiculous to many people. If your intention is to offer me some evidence of your support, you have already done that, and much more effectively, in a way that touched me, by giving me the means to work without worry."

Although Pierre was content, he earned only 300 francs per month, about the same salary as a factory laborer. Now Marie's presence in his life provided the impetus to establish himself professionally, beginning with the preparation of a doctoral dissertation that described his findings of the previous four years, under the title "Magnetic Properties of Bodies at Diverse Temperatures."

Marie watched with interest as Pierre successfully defended his thesis at the Sorbonne on a March afternoon in 1895, before a jury of faculty members that included her physics professor, Gabriel Lippmann. These men held Pierre's future in their hands, but even as they sat in judgment of him, they assumed postures of rapt attention to his expert presentation. "I remember the simplicity and the clarity of the exposition," she later wrote, "the esteem indicated by the attitude of the professors, and the conversation between them and the candidate, which reminded one of a meeting of the Physics Society."

On the strength of his new credentials, as well as recommendations from distinguished scientists, Pierre rose to a new position created expressly for him as professor in the school where he had already worked for twelve years, the *École Municipale de Physique et de Chimie Industrielles*. His pay nearly doubled, to 6,000 francs per year. Nevertheless he declared himself willing to relocate to Poland if that was what Marie required of him.

In mid-July, she told her brother the reason for her abrupt reversal of plans regarding the summer holiday. She would not be returning to Warsaw as usual—perhaps not ever. Józef replied:

I think you are right to follow your heart, and no just person can reproach you for it. Knowing you, I am convinced that you will remain Polish with all your soul, and also that you will never cease to be part of our family . . . And we, too, will never cease to love you and to consider you ours.

I would infinitely rather see you in Paris, happy and contented, than back again in our country, broken by the sacrifice of a whole life and victim of a too-subtle conception of your duty. What we must do now is try to see each other as often as possible, in spite of everything.

A thousand kisses, dear Manya; and again let me wish you happiness, joy and success. Give my affectionate regards to your fiancé. Tell him that I welcome him as a future member of our family and that I offer him my friendship and sympathy without reserve. I hope that he will also give me his friendship and esteem."

Józef and his young family could not travel to the wedding on July 26, 1895, but Marie's father and her sister Helena came, and of course Bronya and Kazimierz Dluski. The ceremony took place at the town hall in Sceaux, where the bride and groom exchanged vows but not rings. Afterward Pierre's parents hosted a small reception in the garden at their home. Then the newlyweds rode off on bicycles to honeymoon among the fishing villages of Brittany.

Marie and Pierre Curie as newlyweds, 1895

"When you receive this letter," Marie told her childhood friend Kazia, "your Manya will have changed her name. I am going to marry the man I told you about last year in Warsaw. It is a sorrow to me to have to stay forever in Paris, but what am I to do? Fate has made us deeply attached to each other and we cannot endure the idea of separating."

She would have written sooner, she apologized, but

she had only very recently, and "quite suddenly," reconciled herself to settling permanently in France.

"When you receive this letter, write to me: Madame Curie, School of Physics and Chemistry, 42 rue Lhomond. That is my name from now on. My husband is a teacher in that school. Next year I shall bring him to Poland so that he will know my country, and I shall not fail to introduce him to my dear little chosen sister, and I shall ask her to love him."

Chapter Three

MADAME CURIE
(Tungsten and Molybdenum)

———

PART OF BECOMING Madame Curie meant learning how to cook. This proved neither difficult nor mysterious, merely time-consuming. Bronya supplied recipes for a few familiar dishes, along with some hands-on demonstrations. Marie also purchased a cookbook and worked through it methodically, making marginal notes on her success or failure in each culinary effort. Pierre, who barely heeded what he ate, discerned no difference between her successes and her failures, so that in this regard, as in other areas of their lives, they seemed ideally matched.

As Pierre's wife, Marie was permitted to move her magnetized steel project into the industrial school where he taught. Pierre still lacked his own designated lab space, even after his promotion to professor, but he continued to set up his experiments in the student laboratories, or, when those were occupied, in the corridor leading from the laboratories to the stairway. Here he carried out his research on the growth of crystals, trying to determine the factors that promoted their development. Here Marie installed the equipment for testing magnets made from various brands of steel.

After working within sight of each other through the day, they linked arms and walked the five blocks home to their sparsely furnished flat on the rue de la Glacière, where they dined at opposite ends of a white wooden table that doubled as a shared desk space.

In the evenings Pierre wrote out the lectures for his new physics course. These alternated at first between crystallography and
electricity. Years before, in 1880, he and his brother, Jacques, had
united these two disciplines by discovering that certain crystals
generated an electric current when pushed or pulled out of shape.
The brothers named the effect piezoelectricity, meaning electricity
due to pressure.

Marie, who planned to become an instructor like Pierre, spent the
after-dinner hours studying for the competitive qualifying exams,
the *agrégation*, which would gain her the necessary certification for
teaching at a girls' secondary school.

"Everything goes well," she assured her brother, Józef, in November.

We are both healthy and life is kind to us. I am arranging my flat
little by little, but I intend to keep it to a style which will give me
no worries and will not require attention, as I have very little help:
a woman who comes for an hour a day to wash the dishes and do
the heavy work. I do the cooking and housekeeping myself.

Every few days we go to Sceaux to see my husband's parents.
This does not interrupt our work; we have two rooms on the first
floor there, with everything we need; we are therefore perfectly
at home and can do all the part of our work that cannot be done
in the laboratory.

When it is fine we go to Sceaux by bicycle; we take the train
only when it is raining cats and dogs.

At the laboratory she had the use of ovens, baths, batteries, galva-
nometers, and other apparatus for putting her steel samples through a
variety of trials. At home or at Sceaux she analyzed her observations,
calculated, interpolated, tabulated, and planned what experiment to
try next in the lab.

One by one she turned each of her forty-seven steel samples into
a magnet with the aid of current-carrying wires. The flowing electricity in the wires created a magnetic field capable of inducing magnetism in susceptible materials. Steel, by its nature, was eminently

susceptible to magnetization. Soon Marie could document which samples—that is, which chemical recipe for steel, as revealed to her by the individual manufacturers—acquired the greatest magnetic strength. She also sought to determine which physical treatments— which specific combinations of heat, cold, and pressure—would best preserve and prolong the much-desired magnetism. For example, some commercial manufacturers subjected their magnetized steel to percussive shocks, which Marie mimicked by intentionally dropping samples to the floor or hitting them with a hammer.

The summer of 1896 found her sitting for the teaching examinations, which lasted several weeks from July through mid-August and coincided with her first wedding anniversary. True to her academic record, she passed in first place. She and Pierre celebrated by taking another bicycle vacation. One evening on their travels south, "lingering until twilight in the gorge of the Truyère," she recalled years later,

> we were enchanted to hear a folk song dying away in the distance, carried to us from a small boat descending the stream. We had taken so little notice of the time that we did not reach our lodging before dawn. At one point on our way back there through the dark we encountered some carts drawn by horses that were frightened of our bicycles, obliging us to cut across the tilled fields. At length we regained our route on the high plateau bathed in the unreal light of the moon. And cows that were passing the night in enclosures came gravely to contemplate us with their great, tranquil eyes.

THE SECOND YEAR of marriage passed much like the first. Marie had her teaching certification now, but no teaching position, and continued her study of steel full-time.

In early March of 1897 she apologized to her girlhood friend Kazia for being late with her annual birthday greetings but gave good reasons for her tardiness. "I am going to have a child," she wrote. "For more than two months I have had continual dizziness, all day long from morning to night. I tire myself out and get steadily weaker, and although I do not look ill, I feel unable to work and am in a very bad state of spirits."

Her weakened condition troubled her all the more because of the worrisome situation in Sceaux. As she explained to Józef, "My husband's mother is still ill, and as it is an incurable disease (cancer of the breast) we are very depressed. I am afraid, above all, that the disease will reach its end at the same time as my pregnancy. If this should happen my poor Pierre will have some very hard weeks to go through."

After the early phase of discomfort and dizziness that she reported to Kazia, she felt well enough to resume her experiments. By July she was ready to apply the final test—the test of time. How long would the "permanent" magnets she had produced by her best methods retain their magnetism? This phase offered a prime opportunity for a restful vacation break. Marie arranged to meet her father at a small hotel on the rocky beach in Brittany's Port Blanc, where Pierre would join them after giving his final lectures of the semester.

It was the first time either of them had spent more than a few hours away from the other since their marriage. Pierre pined for her in his best beginner's Polish: "My little girl, so dear, so sweet, whom I love so much, I had your letter today and was very happy."

"My dear husband!" she replied by postcard, mindful of his limited vocabulary in her language. "Fine weather here, the sun is shining and it's warm. I'm very sad without you. Come quickly. I wait for you from morning till evening and still don't see you coming. I'm fine. I work as much as I can, but Poincaré's book is more difficult than I expected. I need to talk with you about it and look together with you at the parts that are giving me trouble."

Delayed by the need to stay in Sceaux with his ailing Maman, Pierre reverted to French in longer letters. "I think of my dearest who fills up my life, and I long to have new powers. It seems to me that in concentrating my mind exclusively on you, as I am doing, that I should succeed in seeing you, and in following what you are doing; and also to make you feel that I am entirely yours at this moment,—but the image does not come."

At last his brother's arrival from Montpelier freed Pierre from his filial duties, and he reached Port Blanc in August. He and Marie, now in her eighth month of pregnancy, set off together for Brest in their usual fashion, by bicycle. They loved "the melancholy coasts of Brittany," as she described them, but on this occasion she lacked the stamina to fully appreciate "the reaches of heather and gorse, stretching to the very points of Finistère, like claws or teeth burying themselves in the water that forever rages at them."

Cutting their ramble short for a hasty return to Paris ahead of schedule, Marie gave birth on September 12. The new parents named their daughter Irène. Pierre splurged on a bottle of champagne and a few telegrams to Warsaw.

Four days after Irène's birth, Marie returned to the lab to check the strength of her magnets. To her satisfaction, she found no detectable change from her early-summer measurements. Ideally she might have continued the trial for a much longer time, but she had enough data in hand to begin writing up her results with specific recommendations.

Before the end of September, as Marie had feared, Pierre's mother died, and the mood of the family shifted from joy to mourning. Sophie-Claire Depouilly Curie was laid to rest in Sceaux, not far from the house where she had raised her two fine boys. She was the first of the Curies to be buried in the Sceaux cemetery, as her husband's people hailed from Alsace in the northeast. Dr. Curie chose a simple headstone for his wife's grave, with her name at the rounded top and room below for his to be added later:

Madame CURIE
née Sophie-Claire
DEPOUILLY
1832-1897

The bereaved did their best to rebalance their lives after the loss. "I am still nursing my little Queen," Marie wrote her father in

November, just after turning thirty, "but lately we have been seriously afraid that I could not continue." The child's weight had been decreasing, so that Irène "looked ill, and was depressed and lifeless. For some days now things have been going better. If she gains weight normally I shall continue to breastfeed her. If not, I shall take a nurse, in spite of the grief this would be to me, and in spite of the expense; I don't want to interfere with my child's development for anything on earth."

Another week went by before Marie took the difficult step of hiring a wetnurse. She clung, however, to the rest of Irène's care—bathing her, dressing her, recording daily notes on her growth, singing her to sleep at night. Old Dr. Curie, who had delivered the child, now doted on her, finding respite from his grief in her every tiny gesture. He offered to move in with Marie and Pierre, to lend a hand in raising his new granddaughter.

Thus supported, Marie prepared a forty-page summary of her steel research. She pointed out that the steels with the highest content of tungsten tended to hold on to their magnetism most tenaciously, and she defined the optimal tungsten content of magnet steels to be 5.5 percent. Varieties of tungsten steel that also contained 1 to 2 percent molybdenum, she added, had performed even better in her tests. She suspected this finding might surprise people in the industry, since no one else had named the element molybdenum as an aid to magnetization. Having specified the ideal chemical composition, she judged the optimal finishing process to include slow-baking the magnets at sixty to seventy degrees Celsius for about two days.

In December she submitted her report to the *Société pour l'Encouragement de l'Industrie Nationale*. The society announced her findings immediately in its monthly bulletin, praising her work as "important," and published the full text of her paper soon afterward. The 1,500 francs she received as compensation for her research enabled Marie to repay a kindness. She well remembered how the 1893 Alexandrovitch Scholarship of 600 rubles had allowed her to live another year in Paris, to work toward a second university degree, to accept a challenge from the steel industry, and to enjoy all that had followed from those experiences.

Here was an opportunity to give someone else a chance at advancement.

Her show of gratitude startled the granting agency and made Marie the anonymous patroness of another deserving student—someone as impoverished as she had been, and whose identity she would never know.

Marie's advice on magnetized steel reappeared in numerous abstracts and translations. While some reviewers faintly praised the work as "patient and systematic," and a few metallurgists objected loudly to a newcomer's intrusion into their field, a summary appraisal issued several years later by British industrialist and inventor Rookes Crompton acknowledged the overall positive effect of her contribution: "All instrument-makers are deeply indebted to Marie Curie for the excellent work she has published in regard to the saturation and persistence of magnetism in steel bars. Madame Curie has pointed out how much depends on the exact temperature to which the magnet steel must be heated before being plunged, and if her directions are closely followed excellent and concordant conditions invariably follow. The work that she has given to the world in this respect is almost unique in its character and accuracy."

Toward the end of the Curie family's eventful year, on December 16, 1897, when Irène was three months old, Marie recorded the first notes regarding the new research direction she had chosen for her doctoral dissertation. She happened to jot them down in a lab notebook of Pierre's. At home she kept two notebooks of her own for recording household expenditures and Irène's progress (often measuring the infant's weight both before and after feedings). To take up the unused pages of her husband's notebook for the new project seemed the expedient thing to do, and not at all unreasonable in the context of their shared existence.

"In our life together," she wrote in retrospect, "it was given to me to know him as he had hoped I might, and to penetrate each day further into his thought. He was as much and much more than all I had dreamed at the time of our union. My admiration of his unusual qualities grew continually; he lived on a plane so rare and so elevated that he sometimes seemed to me a being unique in his freedom from all vanity and from the littlenesses that one discovers in oneself and in others, and which one judges with indulgence although aspiring to a more perfect ideal."

Chapter Four

PIERRE (Uranium)

FOR HER DOCTORAL research, Marie turned away from magnetism, choosing instead to pursue the startling new energy exuded by uranium. So-called "uranic rays" had been discovered early the previous year, 1896, by the eminent Parisian physicist Henri Becquerel of the nearby natural history institution *Le Muséum d'Histoire Naturelle*. Becquerel's finding had thus far failed to draw much attention from other physicists, because his uranic rays paled in comparison with a prior discovery, by Wilhelm Roentgen in Bavaria late in 1895, of "X-rays" that could pierce human flesh to expose the bones and organs.

The uranic rays' relative lack of popularity among scientists made them all the more attractive as a dissertation topic for a novice such as Marie. While more than a thousand scientific papers had already been published about X-rays, she faced little or no competition on the topic of Becquerel's uranic rays. Moreover, she could make the measurements she intended using instruments invented by her husband and his brother.

Becquerel had come upon uranic rays by accident, finding one thing while looking for another. He suspected that sunlight might elicit the newfound X-rays from certain natural materials, and he conceived an experiment to separate the effects of visible light from those of the invisible X-radiation. First he double-wrapped a photographic plate in heavy black paper. Atop this base he set a sample of a uranium salt called potassium uranyl sulfate—a crystalline compound with the ability to fluoresce (that is, to absorb light and emit its own glow in response). Becquerel planned to expose the wrapped plate and topping of salt crystals to the sunshine of broad daylight. The crystals would glow, and if that glow contained X-rays, then the X-rays would penetrate the black paper—as visible light could not—to impress an image of the crystals on the photographic plate.

As it happened, the February weather proved dreary, and the lack of sunlight compelled Becquerel to stow the wrapped plate and fluorescent material in a drawer while waiting for the skies to clear. Days later, under a still-lingering cloud cover, he retrieved the materials and, perhaps in frustration, developed the plate without having carried out the planned experiment. To his surprise, the plate revealed a vague image formed by the crystals that had been its companions in the drawer. There in the darkness, with no assistance from the sun or any other light source, the uranium salt had somehow transmitted a signal through the black wrapping, into the photographic emulsion.

Follow-up experiments convinced Becquerel that every uranium-containing substance available in the museum's collections could work this same magic. Uranic rays not only penetrated black paper but also passed almost as easily through coverings of cardboard, aluminum, copper, and platinum. Stranger still, the effect did not diminish with time. Month after month, the uranic materials retained their ability to release uranic rays.

Becquerel tried a few other experiments with uranic rays before giving them up. He knew that several investigators across Europe had shown X-rays to be capable of draining electric charge—of "discharging" any electrified bodies in their path. A glass rod, for example, that had become electrified by being rubbed with a cloth would lose its charge in the presence of X-rays. Becquerel found that uranic rays were likewise capable of discharging, though he could not explain the effect.

It fell to physicist J. J. Thomson of the Cavendish Laboratory in England to align these findings, in 1897, with a new theory of the atom. Thomson's atom was electrically neutral overall, comprising subatomic bits of equal but opposite charge, which he named "ions." X-rays or uranic rays traveling through the air split the individual

SCHOOL OF PHYSICS AND CHEMISTRY   37

atoms of atmospheric gases, separating the positive ions from the negative ones. Thus "ionized," or fragmented into ion pairs, the air became a conductor of electricity, through which charge drained away.

Mme. Curie saw in ionization a way to quantify the uranic rays released from various substances—by measuring the electrical conductivity they excited in the air around them.

At Pierre's request, the director of his school ceded Marie a small storage room on the ground floor to use as her laboratory. The space was dank and unheated, but it was hers. She spent the first six weeks in her new niche setting up, trying out, and mastering her experimental apparatus. It consisted of a series of connected devices, both homemade and commercial.

At one end of a worktable, she placed the small ionization-testing chamber that she and Pierre had built of wood from grocery crates. Inside the chamber she mounted two metal discs, each eight centimeters (a little over three inches) in diameter, one above the other, just three centimeters (about an inch) apart. She attached the lower disc to a hundred-volt battery, and the upper one to an electrometer that gauged electric current in units named for French physicist André-Marie Ampère. The Curie piezoelectric quartz stood close by on its tripod, and was also hooked up to the electrometer. As Pierre and Jacques had shown in their pioneering work on piezoelectricity, a current could be elicited from the long, narrow quartz crystal in response to pressure, such as the stress of pulling on it.

To make a measurement, she sprinkled a thin layer of powdered uranium-containing material on the lower disc in the ionization chamber and flipped a switch closing the circuit between the disc and the battery. As the uranic rays ionized the air between the discs, current flowed from the charged bottom plate to the top one, at which point the electrometer responded, its needle rising from zero. Now Marie dropped a tiny weight into the balance pan suspended from the piezoelectric crystal, thus stretching it and causing it to generate an opposing current that nudged the needle back to zero. She needed to keep her eyes riveted on the readout and her right hand with its fistful of weights poised over the balance pan. When the current reached its peak value, or saturation point, and the crystal bore just enough weight to compensate, she clicked off the stopwatch in her left hand. The stronger the activity of the sample, the faster the current in the chamber reached saturation. Later she could calculate the current's strength, in millionths of an ampere, from the weight that matched it, in grams.

She practiced the procedural steps repeatedly to achieve a smooth coordination between clocking and counterbalancing, but she could do little to control conditions in her laboratory. On February 6, 1898, she recorded a near-freezing room temperature of 6.25 degrees Celsius, which she emphasized in the lab notebook with ten exclamation points. The cold and damp affected the instruments nearly as much as they decreased her personal comfort, but she soon established the activity of pure uranium as a basis of comparison for all her other assessments. Patient testing proved that different compounds of uranium produced uranic rays in direct proportion to the amount of uranium they contained. The uranic-ray activity did not disappear or even dissipate when uranium combined chemically with other elements. Nor did the activity of her samples alter if she heated them to high temperatures, or exposed them to strong light, or bombarded them with X-rays. Nothing sapped uranium's emissive power. On the basis of these observations, she concluded that the release of uranic rays must be an essential atomic property of uranium, as constant and defining as its atomic weight.

Yet uranium's behavior defied the most fundamental physical principles. Basic "laws" of physics stipulated that energy could be neither created nor destroyed, but only changed from one form to another. What form of energy existed inside an atom of uranium to generate uranic rays? Was the emissive power the province of uranium alone? She began testing other materials to find out. Gold and copper, she noted, showed "no rays."

At home, with similar diligence, she continued to record Irène's progress. That February, for example, the five-month-old child, who could already "change her position on the bed by rolling" and "hold objects in her hand," suddenly became "afraid of strangers and unfamiliar objects, loud voices, etc."

When Marie had worked through all the substances she had on hand at the school, she borrowed others from fellow scientists. A uranium-rich ore known as pitchblende gave her a jolt when it registered more activity than pure uranium. The odd result made her repeat a number of prior measurements. As she had come to expect, the other uranium-containing compounds consistently evinced less activity than pure uranium. But again and again, and altogether contrary to her expectations, the pitchblende gave off more radiation than its content of uranium could explain.

As she tried to make sense of those results, she continued testing other materials. Soon she found that the element thorium, another heavy metal like uranium, emitted the same sort of spontaneous radiation. This meant that uranium was not the sole source of uranic rays. Indeed, she now suspected that pitchblende owed its strikingly high activity to some as yet unknown element hidden in the mix of its ingredients.

Pierre put aside his crystal-growth experiments to join Marie on the promising new tack her research had taken. Together they filled the remaining pages of the shared notebook with numbers and lists summarizing her findings, and with a graph comparing the density and composition of all the uranium- and thorium-containing compounds. In mid-March 1898 they started a new lab notebook in their joint pursuit of pitchblende's active component. On March 31, at home, Marie documented the discovery of Irène's first tooth.

The work Marie had accomplished to date merited presentation to the *Académie des Sciences*, the authoritative and influential body to which Henri Becquerel belonged. Marie could not deliver the report herself, because she did not belong to the *Académie*. No woman had ever been elected to membership. Nor had Pierre attained the lofty status of an *Académicien*. Marie turned for help to her former professor, Gabriel Lippmann, who read aloud her paper,

"Rays Emitted by the Compounds of Uranium and Thorium," at the April 12 meeting.

The most stunning remarks in Mme. Curie's paper concerned the mineral pitchblende. She had examined three samples of the blackish ore, two from mines in eastern Europe and one from Cornwall in England. Each had yielded a different activity reading, with one of them tripling and another quadrupling the value for uranium. "This fact is very remarkable," she affirmed, "and leads to the belief that these minerals may contain an element which is much more active than uranium."

Now all she had to do was find it.

———

IN THE GROUND, a chunk of blackish-brown pitchblende rock had a dull, greasy look, akin to bubbling tar. The ore was mined for its chief component, uranium oxide, which found wide commercial use as a pigment for coloring glass and pottery a greenish shade of yellow. The means for extracting uranium oxide were already well known and long practiced by the time Marie Curie took an interest in pitchblende. But she was seeking a trace element, perhaps only 1 percent of the ore by weight. She had stumbled upon the unknown substance in her makeshift laboratory, using novel techniques that formed no part of the traditional chemist's or prospector's tool kit. Indeed, the sole known property of her supposed new element was its active emission of uranic rays.

As physicists, neither Pierre nor Marie possessed the knowledge and experience to chemically dissect pitchblende and identify the unknown radiation source it harbored. What they lacked in background, however, they more than made up for in motivation and advice from colleagues. They obtained a hundred-gram (quarter-pound) lump of pitchblende, pulverized it, and began attacking the powder with acids to break it down.

At each stage of their chemical assay, the Curies tested the breakdown products for uranic rays. After the initial acid attack, for example, they treated the solution with hydrogen sulfide, a colorless gas that smelled like rotten eggs; the uranium and thorium remained in the solution, but other substances reacted with the sulfur and fell to the bottom of the beaker as a solid precipitate that proved very active. It contained familiar elements—lead, bismuth, copper, arsenic, and antimony, all recognizable by their behavior in the reactions—and presumably also the mystery element, because Marie had already tested all the known elements, and none of them could account for the electrical effects that the precipitate excited in the ionization chamber.

Next they dissolved the precipitate in ammonium sulfide, and this time the arsenic and antimony stayed in solution, while everything else precipitated out. Continuing in careful stages with different reagents, they arrived at last at a smidgen of residue that behaved chemically like bismuth—a whitish metal similar to lead—except for the fact that it emitted the telltale rays. It was four hundred times more active than pure uranium.

Henri Becquerel, who had listened attentively to the reading of Marie's April report, visited the Curies several times that spring at their lab in the industrial school. In July, when they were ready to release the next outcome of their research, he represented them at the *Académie*.

Marie had taken to calling uranic rays by the more general term "Becquerel rays" after she detected them in thorium. Now she introduced an altogether original term for the ability of select heavy elements such as uranium and thorium to radiate: "radio-active" appeared for the first time in the title of the report she coauthored with Pierre, "On a New Radio-active Substance Contained in Pitchblende," which Becquerel read to the assembled *Académiciens* on July 18, 1898.

The Curies admitted they could not yet separate their new radio-active substance from bismuth by any means. Nevertheless, they felt so certain of the element's existence that they had already christened it: "We propose to call it *polonium* from the name of the country of origin of one of us."

Marie mailed a copy of the discovery report to her cousin Józef Boguski at the Museum of Industry and Agriculture, who saw to its immediate publication in a monthly Warsaw magazine. At the same time, she received an incentive award of nearly 4,000 francs from the *Académie des Sciences*, which greatly enriched the Curies' research fund. Although Marie had won the *Prix Gegner* in recognition of her work with magnetized steel and of her recent investigation into uranic rays, word of the honor reached her indirectly, via letters to Pierre. "I congratulate you most sincerely," said one from an *Académie* official, "and beg you to present my respectful compliments to your wife." Even the award certificate, though it had "Madame Curie" penned in at the top, addressed the honoree twice as "Monsieur."

The Curies spent the summer holiday of 1898 in a small house they rented in the Auvergne region in south-central France. From their base at Auroux, they made bicycle tours to the surrounding towns, explored the hills and grottoes dotting the volcanic landscape, and introduced Irène to the pleasures of skinny-dipping. "For the past three days we have bathed her in the river," Marie noted in her private journal on August 15.

When they returned to Paris, they ordered more of the expensive pitchblende ore from the St. Joachimsthal mine in Bohemia. Pitchblende had shown itself to be such a medley of materials that it merited further analysis. A chemist at Pierre's school, Gustave Bémont, assisted the couple in systematically deconstructing the second quantity of pitchblende, leading to the discovery of a second new radio-active element. In the same way that polonium had adhered to bismuth—signaling its individuality by radio-activity alone—the Curies' second find clung to barium. And Marie had already shown that pure barium lacked any hint of radio-activity.

Once again, the property of radio-activity indicated the presence of an otherwise undetectable element.

Before announcing their result publicly, the Curies sought corroborating evidence from the field of spectroscopy. The technique of spectrum analysis, developed in the 1860s, gave chemists the means to identify elements by the color of light they emitted when heated to incandescence. Each element proclaimed its presence through one or more wavelengths of light, in a pattern as distinctive as a fingerprint. A few elements that had been discovered by means of spectroscopy bore the names of the colors that revealed them, such as cesium (sky blue), rubidium (ruby red), and thallium (from the Greek word for a fresh green twig). The Curies had offered their quantum of polonium to the well-known spectroscopist Eugène Demarçay, who, unfortunately, had failed to find any spectral lines aside from those of bismuth in the tiny sample. They returned to Demarçay now, full of hope, and this time he detected a line in the near ultraviolet belonging to no known element. Better yet, the intensity of the line increased or decreased according to the level of radio-activity in each of the several specimens he was given to examine.

The new element practically named itself—not by any color but rather by its extraordinary degree of radio-activity, which multiplied that of uranium a thousandfold. The Curies and their collaborator Bémont thought the difference might be even greater, but they had run out of pitchblende and could go no further till they got more. On the day after Christmas in 1898, Henri Becquerel informed the *Académie des Sciences* of the discovery of "radium."

Although Becquerel shared the Curies' excitement, the news of the new element elicited little response outside a small circle of physicists.

WITHIN HER FAMILY circle in the five months between the announcements of polonium and radium, Marie lost the ever-ready companionship of her sister Bronya. The doctors Dluski moved in the autumn of 1898 to Zakopane, part of Austrian Poland in the Tatra Mountains, to create a modern tuberculosis sanitarium.

"You can't imagine what a hole you have made in my life," Marie wailed to Bronya in early December. "With you two, I have lost everything I clung to in Paris except my husband and child. It seems to me that Paris no longer exists, aside from our lodging and the school where we work."

As she pressed forward at her laboratory over the following months, and chased after the toddler who could now walk unassisted, she wrote to Bronya in more even tones: "I miss my family enormously, above all you, my dears, and Father. I often think of my isolation with grief. I cannot complain of anything else, for our health is not bad, the child is growing well, and I have the best husband one could dream of; I could never have imagined finding one like him. He is a true gift of heaven, and the more we live together the more we love each other."

Chapter Five

ANDRÉ (Actinium)

———

BY THE CURIES' count, there were now four "radioelements"—uranium, thorium, polonium, and radium. Most other scientists, however, recognized only the first two.

The mere wisps of polonium and radium that the Curies had pried from pitchblende were too minute for the weight calculations or other experiments that would bolster any claim to element status. They had nothing but their radioactivity to show for themselves, plus one lone spectral line. And yet, on photographic plates—the medium in which uranic rays had first made their presence known—the Curies found that polonium and radium could produce in half a minute the same effects that uranium and thorium required hours to achieve.

"In our opinion," Marie wrote in a reflection on her joint work with Pierre, "there could be no doubt of the existence of these new elements, but to make chemists admit their existence, it was necessary to isolate them." The couple would need much larger supplies of raw material for this next phase. Marie's early estimate of a 1 percent polonium content in pitchblende had been wildly optimistic. Since sizable lumps of pitchblende had yielded only barely discernible traces of the desired end products, the proportion by weight veered toward one-thousandth or even one-hundred-thousandth of one percent. And so, instead of a mere hundred grams of pitchblende, they now ordered one hundred kilograms (about 220 pounds) from the mine in Bohemia and paid for it with Marie's prize money.

Clearly their operations would no longer fit into the tiny lab space on the ground floor of the industrial school. Once again the accommodating director, Paul Schützenberger, found room for Marie's activities. Across the courtyard from the school stood an old wooden shed that had once served as an anatomy theater for medical students. A cast-iron stove, a blackboard, and a few worn pine tables were the only furnishings left in the place, which looked more like a hangar than a laboratory. It had an asphalt floor and a glass roof that leaked.

A century would pass before the term "glass ceiling" gained currency as a metaphor for invisible barriers to women's advancement, but Marie Curie toiled under an actual glass ceiling from 1899 to 1902, the years she spent in that "poor, shabby hangar," spinning pitchblende into radium.

Because the shed lacked ventilation hoods for carrying off poisonous gases, the Curies handled the first stages of the chemical

The Curies' hangar in the courtyard on the rue L'homond, 1899

treatments outdoors in the courtyard, where natural breezes did the ventilating for them. When rain drove them indoors, they continued their extraction procedures, "leaving the windows open."

The prohibitive purchase price of pitchblende might have stymied them, but they soon hit on a way to economize: In place of the raw ore, which contained the commercially valuable uranium, they would settle for mine leavings already picked clean of uranium. A sympathetic colleague in the Vienna Academy of Sciences entreated the Austrian government on the Curies' behalf. The state-run uranium mine in Bohemia, which had been dumping its tailings in the nearby woods, agreed to ship the French scientists as much of the worthless stuff as they wanted, charging them only the cost of transport.

In the spring of 1899, a load of depleted pitchblende ore traveled to Paris by rail freight, covering the final leg from the Gare du Nord to the industrial school on a coal-delivery wagon. As soon as the shipment arrived, Marie rushed outside to cut open one of the big burlap sacks and dig her hands into the dusty brown powder. She found it full of pine needles from the forest floor where the waste had lain discarded.

At the mine the ore had been crushed, then roasted with carbonate of soda, washed with warm water, and bathed in sulfuric acid to capture the desired uranium in solution. The residue relegated to the Curies thus came somewhat predigested, but not really prepared for the extraction of radium. Everything they had done on a small scale in the school's laboratories now needed to be scaled up. "I had to treat as many as twenty kilograms of material at a time," Marie reported, "so that the hangar was filled with great vessels full of precipitates and of liquids. It was exhausting work to move the containers about, to transfer the liquids, and to stir for hours at a time, with an iron bar, the boiling material in the cast-iron basin." Nights found her "broken with fatigue."

She interspersed the brute effort of breaking down pitchblende with the more delicate, less physically taxing work of fractional crystallization—repeatedly dissolving and distilling residues to achieve ever-higher concentrations of radium, which made its presence known by demonstrating radioactivity in the ionization test chamber. "The very delicate operations of the last crystallizations were exceedingly difficult to carry out in that laboratory," she said, "where it was impossible to find protection from the iron and coal dust."

Early on, Pierre had confided to Marie that he hoped their new elements would display beautiful colors. Now, as though in answer to his wish, the beakers of solutions and crucibles of crystalline precipitates surrounded themselves in haloes of soft bluish light. Their glow held the couple spellbound. Many an evening they returned to the shed after dinner to gaze at "our precious products" arranged in flasks and crucibles on tables and boards. "From all sides we could see their feebly luminous silhouettes, and these gleamings, which seemed suspended in the darkness like faint fairy lights, stirred us with ever new emotion and enchantment."

By the light of such alluring illumination, they came to realize the true length of the road ahead: They would need several *tons* of mine waste to arrive at a weighable quantity of either of their new elements. Pierre thought the task too onerous and perhaps even unnecessary. He preferred to turn all his attention to ascertaining the properties of radium, while Marie continued pouring her effort into accumulating more radium—a process that required fewer chemical reactions than the isolation of polonium.

New aid came to them in July 1899 in the person of André Debierne, a young chemist who had attended Pierre's school. Debierne

Marie and Pierre Curie in their laboratory

now worked as a lab assistant at the Sorbonne, just as Pierre had done early in his career. The Curies challenged him to raise the techniques they had pioneered to a quasi-industrial level of production. At the same time they forged a new deal with the organization that manufactured and sold Pierre's scientific instruments, the *Société Centrale des Produits Chimiques*. The *Société* would take over the bulk of preliminary chemical treatment, as directed by Debierne, in exchange for some of the fruits of the Curies' labor—that is, a share of the radium.

LATER THAT SUMMER Marie showed her husband and daughter the splendors of her native country. The three of them got together with her father and the families of all her siblings on the picturesque mountainside near Zakopane where Bronya and Kazimierz were building their new sanitarium. At this Sklodowski reunion, Pierre endeared himself to his in-laws by joining their conversations in halting Polish.

By autumn, an entire ton of pitchblende waste had been preprocessed by the staff at the *Société Centrale*. Debierne, in addition to overseeing this phase of the work, simultaneously obeyed the Curies' directive to probe the ore for new radioelements, as they had done, by trying different breakdown procedures and testing for radioactivity at every stage. He claimed his own find in October 1899. Since the Latin word for "ray" had already been appropriated by radium, the proud discoverer borrowed the Greek equivalent to form the name "actinium."

Although Debierne continued to work in his lab at the Sorbonne, he became a frequent presence in the Curies' shed, and also at their new home. In early 1900 they moved to a rented stucco house on the boulevard Kellerman, where Pierre's father tended a garden, and friends gathered on Sundays for casual scientific discussions. Irène learned to call Debierne "Uncle André."

The Curies' stepped-up production practices put them in the unique position of being able to lend out radioactive materials to other scientists, including Henri Becquerel and several physicists in England, Germany, and Austria, who had read of radium in the *Comptes rendus*, the weekly journal of the *Académie des Sciences*, and initiated their own experiments in radioactivity. Appreciative notes from these grateful recipients agreed that the quality of the Curies' samples promised the best research outcomes, the greatest likelihood of observing otherwise undetectable phenomena.

One day Becquerel absently tucked a sealed glass tube of active material into his vest pocket and developed a burn on his torso in the shape of the tube. The incident, he said, slightly tainted his love for radium. Pierre then intentionally exposed the skin on his own arm, and documented the wound's slow healing process in a report published in the *Comptes rendus*. Both he and Mme. Curie, he noted, often found that the palms of their hands flaked and peeled in response to handling radioactive products, and the tips of their fingers hardened painfully for weeks or months at a time. These discomforts did not worry them or deter them from pursuing their science. In fact, their reported skin lesions aroused the interest of medical doctors, who now looked to radioactivity as a potential palliative for certain dermatologic diseases and even as a means of destroying cancerous tumors.

Excitement surrounding the radioelements bubbled over at the first International Congress on Physics, held in Paris in the summer of 1900 to coincide with the world's fair. On the afternoon of August 8, in an amphitheater at the natural history museum, Pierre summarized all that he and Marie had accomplished to date, while also acknowledging the work of others whose insights and projects were expanding the general understanding of radioactivity. Ernest Rutherford, for example, a young New Zealander studying under J. J. Thomson at the Cavendish Laboratory in Cambridge, discovered through experiments with uranium that uranic rays were of two distinct types. The ones he dubbed "alpha rays" played the primary role in ionizing gases, but they did not travel far and could be blocked altogether by a sheet of paper. The "beta rays," on the other hand, ionized only weakly but were extremely penetrating. They could pass through metal screens and cover considerable distances. Adherents of the new science had also learned that radioactivity was contagious: an active source such as radium could render other objects radioactive, too, at least temporarily. This "induced radioactivity," Pierre said, had contaminated just about everything in the Curies' lab.

Before closing, Pierre stressed the gross disproportion between the infinitesimal quantities of the new radioelements and their enormous activity. He and Mme. Curie, by the chemical methods they employed, had processed several tons of pitchblende to retrieve only a few fractions of a gram of each active material. These products, though still not fully isolated, already showed themselves to be at least one hundred thousand times more radioactive than pure uranium.

———

ALTHOUGH THE CURIES' professional standing and reputation were rising rapidly, their income barely covered their expenses. In 1900 the University of Geneva tempted Pierre with a lucrative physics professorship, also promising him a real laboratory and an adjunct research position for Marie. He accepted, then rejected the Geneva offer, opting instead to remain in France and earn extra money by teaching a course in physics for medical students at the Sorbonne. Marie, too, took on paying work. Four years after earning her teaching certificate, she joined the faculty at Paris's best girls' school for aspiring instructors, the *École Normale Supérieure d'enseignement secondaire de jeunes filles* at Sèvres. She described her pupils as "girls of about twenty years who had entered the school after severe examination and had still to work very seriously to meet the requirements" of the *lycées* where they hoped to teach.

Marie's students initially tittered among themselves and mocked her Polish accent, but they soon came to anticipate her classes with excitement. "We watched from our windows for the arrival of the professor," one alumna remembered, "and as soon as we saw her little grey dress at the end of the *allée* of chestnut trees we ran to take our seats in the conference room."

Traveling to the Sèvres campus several times a week by tram took Marie away from the hangar and slowed her progress toward the determination of radium's atomic weight.

At the time of the announcement of radium, in 1898, the Curies' most active sample consisted mainly of barium chloride, with a soupçon of radium too small to be weighed. Since then, Marie had been assiduously amassing more and more radium by sifting additional ton-quantities of pitchblende residue sent from the mine in Bohemia. In March 1902, toward the end of her second year of teaching, she finally attained enough material to attempt the atomic weight measurement that would establish radium as a bona fide element in the eyes of Mendeleev and other chemists.

Her hard-won treasure was a single centigram of pure radium chloride. Spectroscopy by Eugène Demarçay assured Marie that her sample contained only radium and chlorine bound together in chemical combination as the final stage of her fractional crystallizations. It was technically a salt, distantly related to common table salt. It boasted a million times the activity of uranium.

As small as the sample was (only one-hundredth of a gram), the quantity of radium it contained was smaller still. To pin a definitive atomic weight on radium, she would have to put the sample through several chemical reactions involving elements of known atomic weight, and then calculate accordingly.

Marie weighed her precious, perfectly pure radium chloride several times on a Curie aperiodic balance (accurate to the twentieth of a milligram) and averaged the results. Then she dissolved the sample together with silver nitrate, and the interaction caused the constituents to change partners, yielding silver chloride and radium nitrate. She dried and weighed the silver chloride (a few times over). Then she reversed the reaction to reconstitute the radium chloride and made sure nothing had been lost in the various manipulations. Knowing the accepted atomic weight of silver to be 107.8, and that of chlorine 35.4, she could now assess how much of her radium chloride's weight belonged to chlorine. The remainder was radium, to which she assigned an atomic weight of 225, plus or minus 1.* The number stood alone, unaccompanied by grams or other units, since

---

* The currently accepted value is 226.0254.

it represented the weight of a single atom—an entity that could not be weighed on any sort of scale. The 225±1 figure, nearly double the atomic weight of barium, offered further gratifying proof of radium's individuality.

In May, Marie communicated this important result to her father. "And now you are in possession of salts of pure radium!" he replied. "If you consider the amount of work that has been spent to obtain it, it is certainly the most costly of chemical elements!" This was true, yet the professor seemed unable to grasp the wider importance of his daughter's finding: "What a pity it is," he continued, "that this work has only theoretical interest." Days later, when a telegram informed Marie that a sudden illness had stricken her father, she hurried to Warsaw but arrived too late to see him alive. She clung to her siblings at the funeral, and in September she went again to Poland to reaffirm the family closeness.

Having demonstrated the reality of radium, she spent the next several months preparing her doctoral thesis, describing her years of "Researches on Radioactive Substances" that had since given rise to a scientific movement. On the day of her dissertation defense, June 25, 1903, she appeared at the Sorbonne students' hall wearing the new black silk-and-wool dress that Bronya, visiting for the occasion, had made her buy. The family—Bronya, Pierre, and old Dr. Curie—sat at the back of the crowded room near a coterie of Sèvres students. Marie had invited these young women to attend, in the hope that their presence would embolden her, and also with the goal of showing them where their own studies might lead.

In addition to her mentor Gabriel Lippmann, the three-member faculty jury seated at the oak table included chemist Henri Moisson, who had supplied her with her earliest samples of uranium, and physicist Edmond Bouty. She stood before them, occasionally augmenting her answers to their questions with an equation or diagram sketched on the chalkboard. At length they congratulated Mme. Curie on her historic success as the first woman in France to receive the PhD degree in physics. Automatically she became the first wife—as well as the first mother—to own that achievement. And although most of the audience remained ignorant of the fact, she was also the first person to defend a dissertation while pregnant.

The baby came early—much too early, in Marie's fifth month of pregnancy, during the family vacation on the Île d'Oléron off the southwest coast.

"I am in such consternation over this accident," she wrote of her miscarriage to Bronya in late August. "I had grown so accustomed to the idea of the child that I am absolutely desperate and cannot be consoled. Write to me, I beg of you, if you think I should blame this on general fatigue—for I must admit that I have not spared my strength." In addition to her research, her teaching, her thesis preparation, her care of Irène, and her household duties, she had traveled to London with Pierre in early June, when he lectured at the Royal Institution. "I had confidence in my constitution, and at present I regret this bitterly, as I have paid dear for it. The child—a little girl—was in good condition and was living. And I had wanted it so badly!"

Months later, the Royal Society awarded the Curies the prestigious Davy Medal, named in honor of chemist Sir Humphry Davy, for their outstanding contributions to the field of chemistry. Pierre went alone to London to collect the gold medallion and the £1,000 that came with it. Marie, still weak after her miscarriage, had developed the grippe and a lingering cough, but, as she assured her brother in December, she was merely anemic, with no signs of tuberculosis.

A few paragraphs into her newsy letter, she casually mentioned another recent accolade: "We have been given half of the Nobel Prize." Only lately established in 1901, the Nobel Prizes reflected their founder Alfred Nobel's own varied interests. A scientist, engineer, entrepreneur, poet, and dramatist, Nobel had willed his fortune to the rewarding of merit in physics, chemistry, medicine, and literature. And because his invention of dynamite had exacerbated the carnage of war, he endowed a separate prize for peacemakers.

The Royal Swedish Academy awarded the first Nobel Prize in Physics to Wilhelm Roentgen for revealing the existence of X-rays. Roentgen had been chosen over Marie and Pierre Curie, whose names were put forward that year by the influential French doctor Charles Bouchard, apparently because of radioactivity's medical potential. The Curies were nominated again in 1902 by two physicists—Gaston Darboux of France and Emil Warburg of Germany—but were again passed over. Early in 1903, Pierre received word from one of the nominators that he alone—and not Marie—would likely share the physics prize with Henri Becquerel for the discovery and study of radioactivity. "This would be a great honor for me," he replied, "however I should very much like to share the honor with Mme. Curie, and for us to be considered jointly, in the same way that we have done our work." He reiterated the details of that work to clarify Marie's role, and at length the physics committee agreed to include her.

Of the three joint winners of the 1903 Nobel Prize in Physics, announced in mid-November, only Henri Becquerel attended the formal ceremonies in Stockholm on December 10, the anniversary of Alfred Nobel's death. Pierre expressed the couple's thanks to the Swedish Academy by letter, but regretted they could not possibly travel at that time. As Marie explained to Józef, "I did not feel strong enough to undertake such a long journey (forty-eight hours without stopping, and more if one stops along the way) in such an inclement season, in a cold country, and without being able to stay there more than three or four days: we could not, without great difficulty, interrupt our courses for a long period."

Chapter Six

EUGÉNIE
(Radiotellurium)

"RADIOACTIVITY," the term Marie had coined in 1898 to describe a phenomenon of interest in a scientific report, appeared in the 1903 Physics Nobel Prize citation and quickly proliferated through the popular press. News outlets could not get enough of the husband-and-wife savants who worked together in a dilapidated shed.

"We are inundated with letters and with visits from photographers and journalists," Marie told her brother in the immediate wake of the prize announcement. "One would like to dig into the ground somewhere to find a little peace . . ." Within a few weeks it became clear that the unwelcome attention, which the Curies hoped would soon subside, was on the rise.

"Always a hubbub," she wrote Józef in mid-February 1904. "People are keeping us from work as much as they can. Now I have decided to be brave and I receive no visitors—but they disturb me just the same. Our life has been altogether spoiled by honors and fame." Pierre seconded her laments, informing a colleague how "journalists and photographers of every country on earth" had "gone so far as to reproduce my daughter's conversations with her nurse and to describe the black-and-white cat we have at home." Autograph hounds, social climbers, "and sometimes even scientists come to see us in the magnificent establishment in the rue Lhomond." German chemist Wilhelm Ostwald visited the Curies' shed and described it as "a cross between a stable and a potato-cellar," adding, "if I had not seen the worktable with the chemical apparatus, I would have thought it a practical joke."

Marie threw away the fan mail that arrived in bulk, but admitted she could not help reading some of it. "There were sonnets and poems on radium, letters from various inventors, letters from spirits, philosophical letters. Yesterday an American wrote to ask if I would allow him to baptize a racehorse with my name."

In the scientific journals, radioactivity rattled the foundations of physics and called concepts such as "atom" and "element" into question. Ernest Rutherford, who had moved from the Cavendish Laboratory to McGill University in Canada, conducted experiments with a new colleague there and proposed that radioactive atoms changed their identity when they emitted their alpha or beta rays. Those rays, Rutherford maintained, were actually subatomic particles. By expelling them, an atom of one type transformed into another type—that is, an element changed itself into a different element.

The transformation or transmutation of elements had long since been dismissed as a delusion of medieval alchemists. A modern scientist of the twentieth century dared not speak of turning lead into gold. Yet Rutherford insisted that Nature itself apparently dabbled in a kind of reverse alchemy. Rather than elevate base metals to the rank of treasure, the natural, spontaneous transmutations of radioactivity wrought disintegration and decay. From all indications, the silvery shimmer of uranium, the luster of thorium—even the luminous blue aura of radium—would eventually turn to lead. In the spaces between uranium and lead on the periodic table, an undeniable instability prevailed.

These disclosures disturbed Dmitri Mendeleev. In formulating his periodic system, Mendeleev had presumed true elements to be immutable. Moreover, he viewed the atom as an indivisible entity with no component parts. Each element, he believed, consisted of its own characteristically shaped atoms of a particular atomic weight, all identical to one another and different from the atoms of every other element. He had welcomed the discovery of radium, and positioned the new element on the revised periodic table he issued in 1903, after

Mme. Curie determined its atomic weight to be 225±1. Radium's placement on the table, below barium, made chemical sense to Mendeleev, because radium—like barium, calcium, and others in that same column—readily combined with elements from the halogen group to form salts such as radium chloride and radium bromide. But he had not yet assigned polonium or actinium to any of the remaining empty spaces and was entertaining second thoughts about radium. "Tell me, please," the great chemist reportedly demanded of a Russian friend, "are there a lot of radium salts in the whole earth? A couple of grams! And on such shaky foundations they want to destroy all our usual conceptions of the nature of substance?"

In the eyes of the world, however, scarcity only enhanced radium's cachet. The element attracted even more attention than its humble discoverers. A new periodical, *Le Radium*, debuted in Paris in January 1904, as a bridge between the findings of science and the public's fascination. An editorial in the first issue urged readers to scour the countryside for untapped mineral deposits, and to submit any promising-looking rocks for analysis. The St. Joachimsthal mine in Bohemia, which had once shipped tons of its supposedly worthless tailings to the Curies, was now loath to part with so much as a shred of residue.

Le Radium's financial backing came from industrial chemist Émile Armet de Lisle, who owned a factory in nearby Nogent-sur-Marne that produced quinine medications from cinchona bark to treat malaria. Recognizing the medical potential of radium, Armet de Lisle moved to expand his business through an arrangement with the Curies. In exchange for their counsel and cooperation, he would take over the bulk extraction of radium salts and pay the couple in product. The new facility opened in April 1904, next door to the quinine works, under the name *Sels de Radium* (Radium Salts). Unlike the *Société Centrale des Produits Chimiques*, which had aided the Curies by adding radium extraction to its varied activities, *Sels de Radium* existed solely for that purpose.

In June, when Pierre and Marie were due in Stockholm to deliver the obligatory Nobel lecture, they judged themselves still too unwell to make the trip, so they appealed for a further delay. Pierre suffered

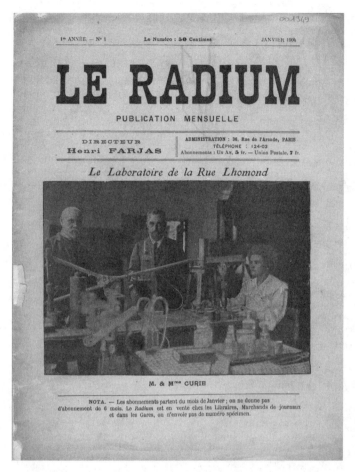

1re ANNÉE. — No 1   Le Numéro : 50 Centimes   JANVIER 1904

LE RADIUM

PUBLICATION MENSUELLE

DIRECTEUR
Henri FARJAS

ADMINISTRATION : 36, Rue de l'Arcade, PARIS
TÉLÉPHONE : 124-03
Abonnements : Un An, 5 fr. — Union Postale, 7 fr.

Le Laboratoire de la Rue Lhomond

M. & Mme CURIE

NOTA. — Les abonnements partent du mois de Janvier ; on ne donne pas
d'abonnement de 6 mois. Le *Radium* est en vente chez les Libraires, Marchands de journaux
et dans les Gares, on n'envoie pas de numéro spécimen.

Cover of the premier issue of *Le Radium* featuring Marie and Pierre
Curie, their colleague G. Bémont, and their piezoelectric quartz

severe pains in his legs and back. Marie was expecting again, and
limiting her activities so as to carry the pregnancy to term. Instead
of traveling to Sweden, they began moving their research projects
from the old shed behind the industrial school to new quarters. The
Nobel Prize had spurred the Sorbonne to create a professorial chair
in general physics for Pierre, along with the promise of a laboratory,
where Marie would preside as *chef de travaux* (chief of operations).
At present the new Curie lab was nothing but an empty chamber of

the Sorbonne Annex in the rue Cuvier, but soon it would include additional rooms.

To conserve her strength, Marie requested a few months' leave from teaching at Sèvres and drafted one of her pet pupils to entertain Irène. The shy six-year-old felt entirely at ease venturing out with Eugénie Feytis, who had become a regular presence at the Curies' Sunday gatherings. Marie was coaching this young Sèvrienne to pursue doctoral-level research in physics, but in the meantime she made an ideal babysitter. One day Eugénie toured Irène through the Sèvres natural history collections, where the child became fascinated by the oversized tooth of an extinct mammoth. She wondered aloud whether Eugénie, who was so much older than she, had ever encountered an actual mammoth. After Eugénie explained she had not been born quite so long ago as that, Irène resolved to question Grandpère about the animal when she got home.

The persistence of material over time preoccupied Marie at this juncture. Polonium, her first-found element, tended to dissipate quickly, within a period of months. She still had not gathered enough of it to detect its spectrum or determine its atomic weight. There was reason to fear polonium might share the fate of its namesake country, which had disappeared from the map of Europe. Already, a scientist in Berlin, Willy Marckwald, had dismissed polonium as a phantom, and was touting his own newfound "radio-tellurium" in its place. Marie suspected that Marckwald's radio-tellurium and her polonium were really the same material, but she would need to prove the point through conclusive experiments.

———

THE CURIES' second daughter, Ève Denise, was born on December 6, 1904. With painful memories of the infant Irène's early weight loss in mind, Marie hired a wetnurse immediately. On February 1 she resumed teaching her classes at Sèvres. "Little Ève sleeps very little," she informed Józef in March, "and protests energetically if I leave her lying awake in her cradle. As I am not a stoic, I carry her in my arms until she grows quiet."

That spring the Curies finally felt healthy enough to go to Stockholm and deliver the long overdue lecture required by the terms of the Nobel Prize. At a special meeting of the Swedish Academy of Sciences on June 6, 1905, Marie sat next to Pierre on the dais while he limned the profound implications of radioactivity.

All radioactive substances, he said, seemed to be "in the course of evolution . . . transforming themselves progressively" from one thing to another. An atom of uranium that issued a uranic ray was no longer uranium but became a different element. Indeed, the fact that uranium and radium always occurred together in nature gave evidence that one gave rise to the other, via a gradual process, over eons.

Pierre took care in his relatively brief remarks to name some twenty other scientists whose investigations had contributed to the current understanding. He cited Marie most frequently, carefully separating the studies she had tackled alone from work they had done together, or that he had pursued either on his own or with different collaborators.

The alpha and beta "rays" originally named by Rutherford had since proven to be particles of vastly different sizes. By their responses to magnetic attraction, they had revealed opposite electric charges. "We have verified, Mme. Curie and I, that the beta rays carry with them negative electricity. The alpha rays . . . behave like projectiles one thousand times heavier and charged with positive electricity." A third type of radioactivity—"gamma rays," more penetrating than the other two combined—had been discovered in 1900 by French physicist Paul Villard, and shown to be electrically neutral.

In addition to these three types of emissions, radioactive elements also gave off heat. Pierre and his protégé Albert Laborde had recently measured the heat released by radium. In just one hour, they concluded, a quantity of radium could raise the temperature of an equal quantity of water from the freezing point to a boil.

Beyond rays, particles, and heat, three of the five known radioactive substances—namely thorium, radium, and actinium—each released an unstable gas. Here Pierre held up a sealed ampoule of such a gas that he had brought from Paris. He said the invisible substance was itself radioactive, undergoing its own transformations at this very moment. And yet the radium that yielded the gas—and emitted an abundance of alpha rays and generated impressive heat—showed no material diminution that could be perceived, even after years of close observation. Polonium, on the other hand, diminished by half within months, "and after several years it has almost completely disappeared."

Alluding to the injury that radium inflicted when held close to the body, Pierre conceded the element might "become very dangerous in criminal hands." He raised the question as to "whether mankind benefits from knowing the secrets of Nature." In answer, Pierre invoked Alfred Nobel's powerful explosives, which had facilitated both "wonderful work" and "also a terrible means of destruction."

"I am one of those who believe with Nobel," Pierre said in closing, "that mankind will derive more good than harm from the new discoveries."

Upon the Curies' return from Sweden, the *Académie des Sciences* elected Pierre a member of its ranks. As an official *Académicien*, he was now entitled to present research results himself, and would not need to lean on Gabriel Lippmann or Henri Becquerel the next time a member of the Curie lab produced noteworthy findings. In addition to Marie and their friend André Debierne, the new group at the Sorbonne Annex included Pierre's collaborator Albert Laborde and two other independent scientists, along with two *préparateurs*, one of whom had followed the Curies from the industrial school, and a couple of graduate students.

Marie kept her focus on polonium. Since coming back to work after Ève's birth, she had been repeatedly observing her polonium to establish the rate of its decline. Experience with the short-lived radioactive gases and their decay products had revealed that each one dissipated by half in a specific period of time, and this "half-value" constituted a defining feature, as characteristic as a pattern of spectral lines. Although four out of the five solid radioelements decayed too slowly to clock their disintegration, polonium seemed hellbent on self-destruction. By Ève's first birthday, Marie could confidently say that polonium dropped to half-value in 140 days. Given that the 140-day half-value for polonium exactly matched the half-value Willy Marckwald assigned to his radio-tellurium, there could be only one conclusion. "Polonium and radio-tellurium are one and the same substance," Marie avowed in the paper that Pierre delivered to the *Académie* on January 29, 1906. Furthermore, she insisted, "it is obviously the name of *polonium* which should be used," in respect of its prior discovery and its landmark status as the first strongly radioactive element found via the novel means devised by "Monsieur Curie and myself."

Marie with Irène and Ève, 1905

THE CURIE FAMILY spent the Easter weekend at a country cottage in Saint-Rémy-les-Chevreuse, just southwest of Paris. It was their current favorite get-away place, replete with meadows and ponds for exploring or resting beside. The parents enjoyed a rare sunny spring day outdoors, watching Irène chase butterflies, laughing together at Ève's antics. Pierre went back to Paris Monday evening, but Marie stayed an extra day for the girls' sake. On Wednesday she joined Pierre in the city for the Physics Society dinner at the Foyot Restaurant. On Thursday everything changed.

Pierre was hurrying through the rain from a meeting of the Association of Professors of the Science Faculties when he stepped into a busy intersection on the rue Dauphine, near the Pont Neuf, and collided with one of the horses pulling a laden twenty-foot wagon. Although the driver later swore he had steered the team to avoid the fallen man, a rear wagon wheel crushed Pierre's skull, and he died there in the street. Calling cards in his pockets told police who he was.

That day, April 19, 1906, which had dawned like any other day, left Marie bereft at day's end, she wept, "with nothing but desolation and despair."

At first she could not bear to tell Irène the truth, so she said only that Pé would not be coming right home because he had hurt his head. When his body was carried into their home that night and laid out in a spare bedroom, Marie found him still warm to her touch. She thought he looked strangely serene, like a man asleep, though she could see the bone of his forehead exposed just below the bandages, and his clothes were soaked in blood.

Jacques came from Montpelier, Józef and Bronya from Warsaw and Zakopane, to guide Marie through the awful early days of her new reality.

On Saturday morning Jacques and André Debierne helped her move Pierre's body into the coffin, and each kissed his cold face for the last time. Marie placed a few periwinkles from the garden beside him, and his favorite photo of her—the one she had sent him during their courtship.

Mourners followed the funeral cortege from the boulevard Kellerman to the small cemetery in Sceaux, where Pierre's name was added to his mother's on the stone marker in the Curie burial plot.

The next week, sitting alone at her workplace in the rue Cuvier, she spilled her thoughts on the graph-paper pages of a lab notebook. "Cher Pierre, whom I'll never see again," she began, "I want to talk to you in the silence of this laboratory, where I could not have imagined I would ever find myself without you." She wrote a dozen pages detailing every moment of their final days together. Even as she strained to render those intimate, idyllic hours indelible in her memory, she could see them blurring and slipping away. "Soon I'll have only your pictures to rely on. Oh! If only I could paint or sculpt a dear vision to keep you ever present in my eyes."

She tried to make a measurement for a graph on which they had each plotted several points, but felt the impossibility of going on. "In the street I walk as if hypnotized, without noticing anything. I will not kill myself. I have no desire for suicide. But among all these vehicles is there not one to make me share the fate of my beloved?"

Over the ensuing weeks, Marie continued to speak to Pierre in this journal of her grief. She cursed the anguish that consumed her. She fretted for her children, especially Irène, "because I dreamed, Pierre, as I often told you, that this girl with her calm, serious manner would one day work alongside you. Now who will give her what you could have given her?"

Meanwhile Pierre's associates at the Sorbonne debated the fate of his laboratory, his vacant chair on the faculty, and the future of his research projects. They settled on his widow as the only person qualified to assume his varied duties. Despite the lack of precedent for inducting a woman into their professorial ranks, the members of the university council made their unanimous decision official on May 13. A few of them blundered awkwardly by congratulating the distraught Mme. Curie on her professional rise, as though she had been promoted simply on merit and not as the consequence of a catastrophic loss.

"I am offered a position as successor to you, my Pierre, your course and the direction of your laboratory. I have accepted. I can't say whether this is good or bad. You often told me you would like me to teach a course at the Sorbonne. And I would like to make an effort, at least, to continue your work. Sometimes I think that in this way it will be easiest for me to live. Other times, I fear I am mad to attempt it."

Part Two

Sorbonne Annex

12 Rue Cuvier

"But was it really uranium?"
"Absolutely: anyone could have seen that. It had an incredible weight, and when you touched it, it was hot. Besides, I still have it at home: I keep it on the terrace in a little shed, a secret, so the kids can't touch it; every so often I show it to my friends, and it's remained hot, it's hot even now."

—Primo Levi, *The Periodic Table*

The man of science must have been sleepy indeed who did not jump from his chair like a scared dog when, in 1898, Mme. Curie threw on his desk the metaphysical bomb she called radium.

—Henry Adams, *The Education of Henry Adams*

Chapter Seven

HARRIET (Emanation)

ON NOVEMBER 5, 1906, the day she was to deliver her first lecture in her new role as the Sorbonne's first female professor, Marie went first to Sceaux, to Pierre's grave. Here, as at the laboratory, she felt close to him. She was thinking seriously of moving to Sceaux, so she could visit the cemetery more often. Relocating outside Paris proper would doubtless be better for the girls, she reasoned, in terms of fresh air, and would also restore her father-in-law to his old familiar neighborhood, with its slower pace and wider grounds for private gardens. The extra distance from rue Cuvier would lengthen her daily commute, but she would find some way to make up the lost time.

A large and curious crowd awaited Mme. Curie in the physics amphitheater at the Sorbonne. Many were fashionably dressed and had never before found reason to be in the building. The new professor's entrance at half past one roused an ovation better fitted to a spectacle than to a lecture. Keeping her eyes down and gripping the long table set with materials for her demonstrations, Marie could see, seated in the front row, the students from her advanced physics class at Sèvres. Although a first-time speaker usually prefaced the inaugural lecture by thanking the university and eulogizing his predecessor, she came immediately to her point.

"When we examine our recent progress in the domain of physics," she began, "a period of time that comprises only a dozen years, we are certainly struck by the evolution of our fundamental notions regarding the nature of electricity and of matter." Only a dozen years—the duration of her life with Pierre. "This evolution happened in part because of detailed research on the electrical conductibility of gases, and also because of the discovery and study of the phenomena of radioactivity."

In 1894, the year she met Pierre, electricity was understood to consist of invisible "cathode rays." These rays could be made visible as dancing streaks of light inside the elongated, evacuated glass tubes produced by British chemist Sir William Crookes. In 1895, the year she married Pierre, a Crookes tube in the lab of German physicist Wilhelm Roentgen caused a plate covered with the fluorescent material zinc sulfide to glow, though the two objects stood on opposite sides of the room. Something had jumped out of the Crookes tube to excite the zinc sulfide, and Roentgen named the something X-rays. In France, Henri Becquerel, while trying to replicate some of Roentgen's experiments, discovered uranic rays in 1896. That same year, J. J. Thomson at the Cavendish Laboratory in Cambridge, along with his student Ernest Rutherford, demonstrated that X-rays split the air into electrically charged ions, allowing electric current to flow. Continuing alone, Rutherford showed that uranic rays did the same.

By the electrical conductivity of air, Mme. Curie had made the precise measurements that led to the discovery of two new elements, polonium and radium, plus a term to describe them: radioactive.

After Rutherford showed radioactivity to be of two types, alpha and beta, Becquerel equated the beta type with the negative ions of electric current, which had since come to be called "electrons." The much larger alpha type carried positive charge. A radioactive atom that ejected an alpha particle instantly transformed itself into a different element. An entire section of the periodic table was thus in constant flux.

As she led her listeners through the events that had shaped her science, the audience members who were neither Sorbonne students nor Sèvriennes quickly lost the thread. Still, they sat silent and attentive, stirred by the sight of the pale figure in the black dress. When she finished speaking, Mme. Curie exited the amphitheater without waiting for another outburst of applause.

"Yesterday I gave the first class replacing my Pierre," she wrote in her private journal. "What grief and what despair!" Addressing Pierre directly, she said: "You would have been happy to see me as a professor at the Sorbonne, and I would have done it so willingly for you.—But to do it in your place, oh my Pierre, could one dream of a thing more cruel? And how I suffered through it, and how discouraged I am." Nevertheless she vowed to do as she had promised, motivated partly by "the desire to prove to the world and especially to myself that the one you loved so much has really some true worth." She no longer felt lively or young. Nothing brought her joy, or even pleasure. "Tomorrow I will be 39," she noted. "I probably have only a little time to realize at least a part of the work I have begun."

In addition to teaching, she had stepped, for the same painful reason, into her new role as laboratory director. Her close friend and colleague André Debierne now filled her prior place as *chef de travaux*. The other salaried workers, the researchers, and the students were adjusting themselves accordingly to the loss of Pierre. Marie made no immediate move to recruit new personnel, but, just before the start of the fall 1906 semester, she agreed to accept a young woman who had once worked as assistant to Ernest Rutherford.

The newcomer, Canadian-born Harriet Brooks, arrived in Paris bringing several years' experience in radioactivity research. She was thirty years old, unemployed for the first time in her life, and fresh from the breakup of her engagement to a fellow physicist at Columbia University in New York. Her first impression of the lab in the rue Cuvier gave her cause for further disappointment.

Harriet located the place with difficulty near the *Jardin des Plantes*, at some distance from the heart of the Sorbonne. Although the scientific equipment appeared adequate to her eye, and the supply of radioactive materials downright enviable, a wide courtyard separated the office part of the complex from the communal workspace, where nine scientists crowded together, practically cheek by jowl. Tales of the Curies' "miserable wooden hangar" had of course reached North America, but that had been their lot before winning the Davy Medal and the Nobel Prize. One would expect the present Curie lab to occupy more convenient, more commodious quarters.

In comparison, the Macdonald Physics Building at McGill University in Montreal, where Harriet Brooks had studied and begun her research career, was large to the point of grandiose. The castle-like edifice included a turret with balconies, and its big rooms boasted the world's finest apparatus. In fact, the acknowledged excellence of the physical plant had lured her mentor, Rutherford, away from the famed Cavendish Laboratory in England. Upon his arrival at McGill in 1898, he chose Harriet Brooks—first in her class and also president of it—as his first graduate student.

The tall, blond Rutherford was only five years older than Harriet, but he acted very much the new professor, all science and all propriety.

Harriet Brooks (back row, center) and Ernest Rutherford (far right) with other physicists at McGill University, ca. 1899

He had a longtime sweetheart back home in his native New Zealand, Mary Newton, whom he planned to marry as soon as he could afford to support a family. Meanwhile he showed true admiration for Harriet's ability, and assigned her a project that proved suitable for her master's thesis: the magnetization and demagnetization of steel needles. She became the first woman to be awarded a master's degree in physics at McGill. This achievement made her an anomaly at home as well. Of the nine Brooks children, only Harriet and her younger sister Elizabeth managed—by striving and by seizing every scholarship opportunity—to attend university.

In autumn of 1900, soon after Rutherford returned to campus from New Zealand as a newlywed, he invited Miss Brooks to assist him with experiments in the new field of radioactivity. Together they demonstrated that radium released some vapor in addition to its alpha rays. This "emanation," as Rutherford named it, was a previously unknown gas, heavier than air, and radioactive in its own right. It seemed to be a new element, though they could hardly believe their evidence. Unlike Mme. Curie's two new elements, which she had extracted from a complex mineral ore, their new gas came directly from radium, as though the one element had birthed the other. Moreover, the new gas went through its own subsequent change, losing its radioactivity over a period of days and leaving a solid residue on the walls of its container. The residue also proved radioactive, though its activity dissipated quickly, within minutes.

On May 23, 1901, Ernest Rutherford and Harriet Brooks reported their "new gas from radium" to the Royal Society of Canada. Mindful that their findings smacked of alchemy, they did not claim to have seen one element turn into another. Nor did they give their new gas a new name. They referred to it simply as "radium emanation"—a product distinct, though not altogether divorced, from radium.

Because McGill did not yet offer a doctoral degree in physics, Harriet continued her graduate studies at Bryn Mawr, the renowned women's college in Pennsylvania. All through the 1901–1902 academic year, she continued to collaborate with Rutherford and to seek his counsel from afar.

"I'm afraid," she wrote him in March 1902, "that your generosity in placing me as a collaborator where I am really nothing more than a humble assistant has rather imposed on the faculty at Bryn Mawr, for last night, they awarded me the European Fellowship." This bounty allowed Harriet to realize the costly dream of research and study abroad. Rutherford helped her secure a place with his own mentor at the Cavendish Laboratory, physicist J. J. Thomson.

While Harriet was in England, Rutherford tapped a confident young chemist at McGill, Frederick Soddy, to wrestle the emanation with him. Their experiments affirmed Rutherford's hunch that radium emanation was indeed a new element. In other words, by giving up an alpha particle, a radium atom changed into an atom of emanation. "This is transmutation," Soddy is said to have blurted out in amazement. "For Mike's sake, Soddy," Rutherford replied in the privacy of the lab, "don't call it *transmutation*. They'll have our heads off as alchemists."

The long-abandoned concept of transmutation, sullied by its associations with mysticism and fraud, now struck Rutherford and Soddy as a natural process with a scientific explanation. Emboldened, they made their ideas public in the summer of 1902. And although Rutherford continued to cite Harriet Brooks in his lectures and writings, only the names Rutherford and Soddy adhered to the breakthrough theory of radioactive transformation—the "new alchemy."

AT THE CAVENDISH, Harriet found everything she needed to continue her work, except for confidence in herself. "I am afraid I am a terrible bungler in research work," she wrote to Rutherford after several months in England; "this is so extremely interesting and I am getting along so slowly and so blunderingly with it." She was trying to compare the emanation released by radium with a similar gas that came from thorium. "I think I shall have to give it up after this year, there are so many other people who can do so much better and in so much less time than I that I do not think my small efforts will ever be missed." Other letters to him mentioned her "deluded moments" and "sorely muddled" work, even as J. J. Thomson commended her "very interesting results" in his own letters to Rutherford.

Harriet learned that Professor Thomson's wife, née Rose Paget, had once been his physics student. After a dozen years of marriage, Mrs. Thomson still showed up at the laboratory every day, but only to prepare the afternoon tea and facilitate conversation between new arrivals and old hands.

Harriet went back to McGill in 1903. Under Rutherford's aegis again, she closely observed the rise of emanation and its decay into substances that coated the chamber walls of her experimental setup or collected on wires she inserted into the chambers. She timed and measured every fluctuation in radioactivity, building up a body of work and publishing her results, till she took a position in 1904 as a physics instructor at Barnard College, the women's arm of Columbia University in New York City.

Harriet had envisioned a teaching career since her undergraduate days, and this particular post carried the added advantage of a reunion with Bergen Davis, the Columbia physicist she had met at the Cavendish. By the summer of 1906 they were ready to announce their engagement. Harriet wanted—expected—to remain on the Barnard faculty after she married, but the dean of the college reacted to her wedding plans by requesting her immediate resignation.

Instead of resigning, Harriet argued in self-defense and against the general unfairness of the judgment: "I think also it is a duty I owe to my profession and to my sex to show that a woman has a right to the practice of her profession and cannot be condemned to abandon it merely because she marries." Unaccustomed as she was to dissent, she apologized for questioning the administration's authority, "but I cannot acquiesce without violating my deepest convictions of my rights."

The dean, backed by the board of trustees, held firm. Miss Brooks could not combine marriage with her college duties because the dual roles would force one of two unsavory outcomes: either she would put her husband before her students, thereby compromising her classes, or she would put her teaching before her husband, making her the sort of wife that the college could not countenance.

Beaten, Harriet agreed to postpone the wedding indefinitely. In August she broke off the engagement. And although marriage had posed no impediment to teaching in her mind, the breakup gave her reason to resign.

"I find that it will be almost impossible for me to fulfill my engagement at Barnard for the coming year," Harriet wrote from a summer address in the Adirondacks, "owing to the unfortunate necessity I have been under, of terminating my engagement to be married. If it is possible I should like to have my place filled and I shall spend the year in study abroad." This resolve carried her to Paris.

Harriet's familiarity with radioactive gases eased her entrée into the Curie lab in the fall of 1906. She partnered with André Debierne to study a third type of emanation—this one coming from actinium, the element he had discovered. Speaking French posed no real problem for her, except that in Paris courtesy required her to use the language, which had been merely optional in Montreal. The surroundings—the banks of flasks, crucibles, and glass tubing blown into custom configurations, the sprawls of gauges, wires, magnets, and miscellaneous hardware—provided familiar ground in a foreign setting. The grieving Mme. Curie, however, could not have contrasted more starkly with the booming, exuberant presence of Rutherford, whose off-key singing often resounded through the halls of the Macdonald Physics Building.

"Physically," Harriet noted, "she is extremely frail and her two children naturally absorb some of her attention." Yet Madame the director proved "ever ready to grapple with the difficulties and problems of those working with her, even when she was a prey to anxieties that would unnerve most women."

Marie's normally quiet demeanor had evolved into an almost total social isolation. In a letter to Eugénie Feytis, her former student who was now teaching physics at Sèvres, she apologized for the change in her behavior. Although her affection for Eugénie had not waned in the slightest, she said, "My life has become so difficult at this point that it's impossible for me to devote any time to socializing. All our mutual friends will tell you that I never see them any more except for professional matters."

She had recently assigned herself the task of redetermining the atomic weight of radium. She possessed much more material now, thanks to the ongoing reciprocal arrangement with Émile Armet de Lisle and his *Sels de Radium* factory. For her 1902 weighing she had made do with nine centigrams of radium chloride; this time she worked with forty decigrams (more than forty times the previous quantity). She had gained more experience, too, though the weighing remained a painstaking affair. Throughout the process the crystals of radium chloride tended to absorb water, necessitating elaborate, repeated drying procedures. In the end she was gratified to see that her 1907 result, 226.2, did not so much change as refine her earlier figure of 225±1.

She was also pleased with Harriet Brooks, and very much wanted her to stay on. Harriet had joined the lab as a *travailleur libre*, or independent researcher, which meant that she received no grant money or guarantee of her position. Funding was as scarce as space was tight. In the spring of 1907, however, the American industrialist Andrew Carnegie gave the Curie laboratory a $50,000 endowment to establish an annual research fellowship. Marie offered the first of these fellowships to Harriet. At the same time, Harriet heard that Rutherford was moving to the University of Manchester, where another fellowship opportunity held out hope of a new place for her at his side.

"Miss Brooks is a very good friend of my wife and myself and I should be delighted if she got the fellowship," Rutherford told his contact at Manchester. "She's at present working with Mme. Curie at the Sorbonne and wishes to work at research in Physics in England, if possible. She is a very able woman with an excellent knowledge of mathematical experimental Physics." He judged her an ideal choice "on her merits, quite apart from personal considerations on my part," and felt "quite confident she has as strong claims as any possible candidate for the position."

Rutherford's official letter of recommendation promoted her even more strongly: "Miss Brooks has a most excellent knowledge of theoretical and experimental Physics and is unusually well qualified to undertake research. Her work on 'Radioactivity' has been of great importance in the analysis of radioactive transformations and next to Mme. Curie she is the most prominent woman physicist in the department of radioactivity." To erase any possible doubt about her capability, he added, "Miss Brooks is an original and careful worker with good experimental powers and I am confident that if appointed she would do most excellent research work in Physics."

In May 1907 Harriet left Paris, still unsure of her future plans. "Dear Madame Curie," she wrote from her London hotel on the last of the month, "I am sorry that I have been so long in letting you know my decision with regard to the scholarship you so kindly offered me for next year." There had been "unexpected delays" regarding the Manchester fellowship, as well as an even more unexpected new development, which she did not explain, except to say she would not be returning to Paris. "I ask you not to keep the appointment in your laboratory open for me any longer."

Rutherford broke the news to the Manchester fellowship committee: "Miss Brooks has just informed me that she is engaged to be married . . . next month." He named her intended as a Mr. Pitcher of Montreal, "an old and persistent admirer" who had followed her to London to propose. "While personally I am sorry not to have her in Manchester, such bolts from the blue are to be expected when ladies are in question."

Frank Henry Pitcher had risen from the position of demonstrator in Rutherford's McGill laboratory to become general manager of the Montreal Water and Power Company. Having reconnected with Harriet shortly before she sailed for Europe, he pursued her via frequent letters over a period of months. In April he mentioned business affairs that required his attention across the Atlantic, giving him a chance to see her again. "Hurry up and discover some quicker way of decomposing or transforming radium," his letter urged her.

They married on July 13, 1907, at the Parish Church of St. Matthew in London, and left immediately for Montreal, with Frank promising his bride that he would "try to make up to you all and more than you think you are now losing."

In choosing him, she chose not to pursue her research. She would write up an account of her recent results for André Debierne as the final act of her professional career.

Marie's altogether different experience of marriage had not only advanced her career instead of ending it, but positioned her to assume her singular role as female head of a laboratory. Wedded to her research partner, she had managed to incorporate love and motherhood into the fullness of a life in science, if only for a while.

"A year has passed," she wrote on the anniversary of Pierre's death. "I live, for your children, for your aged father. The grief is dull but always there. The burden weighs heavily on my shoulders. How sweet will it be to go to sleep and not wake up? How young my poor dear ones are! How tired I feel!"

Chapter Eight

ELLEN

(Copper and Lithium)

———

IN THE LARGE garden at the new home that Marie rented in Sceaux, she found room to install a tall crossbar with a trapeze, flying rings, and a slippery cord for climbing. Ève was still too little to hoist or swing herself, but nine-year-old Irène took to gymnastics immediately, as she did to all outdoor activities, from hiking and bicycling to gardening. She had her own plot of earth now, to cultivate as she chose, with just a little guidance from Grandpère.

"My dear good Mé," Irène wrote home to her mother in the summer of 1907 from a vacation on the Normandy coast with her Uncle Jacques's family. "I am very happy to be at the sea and I am making some very beautiful forts in the sand and some very beautiful scratches, scrapes and grazes on my arms and legs."

Mé and Grandpère concurred that Irène was the kind of child Pierre had been—unsuited to instruction by lecture or long hours of confinement in a classroom. The young Pierre had rambled the woods of the Paris environs and studied with tutors, including his father and older brother, until he entered university. Marie, recalling the restrictive atmosphere of her own primary education, now organized a cooperative school to meet her daughter's perceived needs.

With the help of friends and colleagues who were also parents, she amassed a stellar faculty. Jean Perrin, formerly the Curies' next-door neighbor on the boulevard Kellerman, volunteered to teach chemistry at his Sorbonne lab one day per week. His wife, Henriette, offered history and French in her living room. The Perrins' two children, Francis and Aline, Irène's longtime playmates, thus became her classmates as well, as did seven other familiar youngsters between the ages of eight and twelve. The cooperative school recalled the Flying University of Marie's youth in that its students moved from place to place to take their classes. They traveled by train to Fonteney-aux-Roses to learn English, German, and geography from Mme. Alice Chavannes, whose husband, Edouard, was professor of Chinese at the Collège de France. Henri Mouton of the Pasteur Institute covered biology. The sculptor Jean Magrou showed the group how to model and draw, and also toured them through the collections in the Louvre. Physicist Paul Langevin, a pupil of Pierre's who succeeded him at the industrial school and also served as Marie's replacement at Sèvres, taught mathematics. Marie herself hosted physics class in the Curie lab on Thursday afternoons. She stressed the importance of measurement, the value of rapid mental arithmetic. She showed the children how to make and graduate a barometer, and they were amazed to see the object they had crafted actually function as intended. With it they could measure air pressure to predict changes in the weather.

"This little company which hardly knows how to read and write has permission to make manipulations, to engage in experiments, to construct apparatus and to try reactions," marveled a gossip columnist who caught wind of the goings-on. "So far, the Sorbonne and the building in the rue Cuvier have not exploded."

Only a few city blocks separated the Sorbonne, where the "little company" learned chemistry, from "the building in the rue Cuvier," which housed the Curie lab. But to Marie, the different street addresses signified the gulf between the proper laboratory that Pierre had always wanted and the meager one allotted him. She considered it her duty to enlarge and improve their once-shared domain for the sake of his memory. At present the lab occupied rooms on either side of the Annex's wide courtyard. A small, free-standing pavilion had been built in the courtyard for the Curies in 1904. Inside the pavilion, the output from Émile Armet de Lisle's *Sels de Radium* factory underwent the delicate final stages of purification, by the process of fractional crystallization, to prepare radium chloride for research purposes.

Marie dutifully completed what she could of the studies interrupted by Pierre's sudden death. Soon all of his published papers would be reissued in a collection of his complete works, edited by his two close friends and former students Charles Chéneveau and Paul Langevin. When Paul asked Marie to write a preface, she put aside the private journal in which she addressed Pierre directly and took the opportunity to extol him publicly.

"Certainly," she wrote, "he had the power to exert a profound influence, not only through his great intelligence but also through his moral integrity and the infinite charm that he radiated, and to which it was difficult to remain indifferent." Pierre's reluctance to publish any but the most elegant, definitive results, she explained, limited the present volume to its six hundred pages. In what had proved to be his final year, her late husband had believed himself on the verge of attaining, at long last, the laboratory of his dreams, peopled by a team of collaborators who shared his passion for research, where he might have pursued his brilliant scientific career through an advanced old age. Alas, she concluded, "the Fates ruled otherwise, and we must bow to their inscrutable judgment."

Marie found a new recipient—a German physicist named Hermann Starke—for the fellowship she had offered Harriet Brooks. The crowded lab could not easily accommodate additional researchers, but Marie heeded a note from a Norwegian chemist named Eyvind Bødtker promoting another potential candidate: "Miss Gleditsch," wrote Bødtker, "is a highly educated and intelligent chemist. She would like to work with you solely out of love of science, not to obtain any kind of degree." He assured Mme. Curie that "once you have gotten to know her, you will do anything for her, and you will not regret that you opened up your laboratory for her." To Miss Gleditsch he offered advice and more, down to the exact wording of her letters to the director, given his familiarity with written French.

After all the arrangements were settled, Bødtker congratulated his protégée: "I am pleased to know that you, after so many years of intense work, mostly for others, will at last get out to study under better conditions than here at home. And I am confident it will turn out that you have chosen the correct branch of chemistry."

Ellen Gleditsch, twenty-seven years old and seemingly the sole radioactivity enthusiast in all of Norway, had yet to encounter her first radioelement. She was trained as a pharmacist and supported herself assisting Bødtker in the organic chemistry laboratory at the Royal Frederick University in Kristiania (as Oslo was then known), where all her students and colleagues were men. Findings from the new world of radioactivity that reached her via foreign journals—such as the *Comptes rendus* and the *Journal of the Chemical Society Transactions*—inspired her to embrace the field as her own. She described what she knew of its brief history and the wonders of radium in an article she wrote for the popular monthly magazine *For Kirke og Kultur* (*For Church and Culture*). To engage in the new science, however, she knew she needed to seek experience abroad, whether with Rutherford in Manchester or at the Cavendish or perhaps in Vienna, where active research was also in progress, or preferably in Paris with her idol, Mme. Curie.

Ellen Gleditsch

Once she arrived at the Curie lab in the autumn of 1907, Ellen Gleditsch quickly earned the trust of her new supervisors by dint of her deft, patient, methodical way of working—skills she had honed in her previous professional experience, and even earlier as the responsible older sister of ten younger siblings.

Under the tutelage of Mme. Curie and André Debierne,

Mlle. Gleditsch learned the fractional crystallization technique pioneered in the old shed. Now it was her turn to augment the lab's supply of radium by this process, prying a jot of the much-desired radium chloride from a matrix of mostly barium chloride.

She began, as instructed, by dissolving a portion of the material in distilled water, boiling the solution, then letting it cool in a covered capsule. Radium chloride, being less soluble than barium chloride, crystallized out of solution more quickly. Beautiful yellow-orange crystals formed at the bottom of the capsule, and registered an appreciable degree of radioactivity when tested in the ionization chamber. But this small and still impure harvest was only the first step of a lengthy procedure. Next Ellen decanted the "supernatant liquid"— the portion above the crystals—and allowed part of it to evaporate, which gave her a second crop of crystals, less radioactive than the first. Then she repeated the whole process with both batches, subjecting each to dissolution, boiling, evaporating, decanting, boiling, and drying, and winding up with four portions of crystals. These she reduced to three by combining the middle two; then she put all three again through the specified series of steps. She continued dividing and repeating the treatment until she arrived at a fraction showing no radioactivity at all, which she could safely discard. Turning to the remainder, she repeated the steps afresh. As she continued in this fashion, systematically rejecting the inactive fractions in each portion, the amount of material gradually diminished in quantity but increased in radioactivity, until she had jettisoned all the barium chloride and was left with a modicum of pure radium chloride.

"I don't understand how anyone dared to trust me with it," Ellen later reflected. "I had between my hands a radium preparation worth 100,000 francs."

———

AS ÉMILE ARMET DE LISLE had foreseen, the medical applications of radium inflated the element's value, and rarity drove its price higher still. To meet demand, he scoured the French countryside— and other countries, too, through a network of prospectors—in search of new sources. As each candidate ore came to light, the Curie lab evaluated it and experimented with means for teasing out the radioelements from a mélange of other materials. If an ore yielded a promising result, as determined by repeated measurements of radioactivity, then the lab would instruct the factory in the specific procedure for extraction—and immediately try to streamline or otherwise improve on that procedure. But nothing could alter the harsh fact of diminishing returns in this business: A ton of rock yielded a pound, at most, of raw salts, from which Mme. Curie—and now, Ellen Gleditsch—would derive only one or two milligrams of pure radium chloride.

Ellen said little more than *bonjour* and *bonsoir* during her first few weeks at the lab, hushed by her lack of fluency in French. When she accepted Mme. Curie's invitation to spend a Sunday afternoon *en famille* at her home in Sceaux, she met her best conversational match in ten-year-old Irène. In time she learned to chat just as amiably with Eugénie Feytis, "Uncle André" Debierne, the Perrins, the Langevins, and other regulars at Sunday get-togethers.

To interest Ellen and also to advance her, Marie suggested a research project regarding a new claim about transmutation from Sir William Ramsay of University College, London.

Ramsay had been knighted and awarded the 1904 Nobel Prize in Chemistry for a series of new element discoveries in the 1890s. All his finds were gases, and all of them unusual in their refusal to react with other elements. He named them for their standoffish behavior, from argon (Greek for "idle") to xenon ("stranger"). Ramsay also showed that helium, an element first detected in the Sun's spectrum and long thought to exist only in the Sun, occurred on Earth, too, and belonged to this same group of inert, or "noble," gases that did not mingle with other elements.

The radioactive emanations of radium, thorium, and actinium, being both gaseous and inert, attracted Ramsay's attention as prospective members of his noble-gases family. In a recent study, he had exposed copper—a nonradioactive element—to radium emanation. The result, he reported, was the transmutation of copper into lithium, the lightest of the metals. Other researchers doubted this outcome, even scoffed at it privately, but the only acceptable way to oppose Ramsay was to repeat what he had done and figure out what he had done wrong. This was the task Marie chose for Ellen's initiation: to partner with her in challenging the published results of a well-established and highly decorated scientist.

"During the course of these experiments," Ellen recalled in a memoir, "I was able to see and appreciate Marie Curie at work on a scientific problem. She was very precise in manipulations, she judged everything which resulted with a lively critical intelligence, and she evaluated the results with perfect lucidity. I saw how much she took the success of an experiment to heart. She was devastated when she realized that the introduction of the emanation hadn't succeeded; when everything went well, she was happy, her eyes were luminous, and a smile transformed her ordinarily sad face."

It took them several months to complete their work on the copper-lithium question. In retracing Ramsay's steps exactly, they said they had indeed seen lithium emerge—but not for the reasons Ramsay supposed. Instead of arising from copper by transmutation, the lithium had more likely leached out of the glass vessels in his laboratory. When they substituted platinum containers for glass ones and ran through the process again, they detected no lithium in the end product. This result, a potential embarrassment for Ramsay, capped the first of Ellen's planned two years in Paris with triumph. In the *Comptes rendus* of August 10, 1908, and in the pages of *Le Radium*, "Mlle. Gleditsch" shared a byline with "Mme. Curie."

———

HAVING TAUGHT HER Sorbonne course for two years in Pierre's place, Marie inherited his official title of university professor in 1908. Her fully credentialed recognition as first female professor set a precedent, not only in France but in all of Europe as well. Between her name and her position, she was now a magnet for aspiring radioactivists. She said yes to six new requests from independent researchers seeking places in the Curie lab.

In September, André Debierne presented her current idea for a laboratory expansion to Émile Armet de Lisle at *Sels de Radium*. The industrialist readily confirmed his support for her plan. He felt "very honored," he said, and professed himself "very happy" to house a permanent extension of her research laboratory in his factory. He would put one of his staff at her disposal immediately, and provide an office for M. Debierne, in addition to lab space. "I will do everything necessary," his letter promised, "to accommodate the new operations under your guidance."

The association of Marie Curie's name with his product gave Armet de Lisle an incalculable advantage in an increasingly competitive field. At the same time, the availability of his product in her laboratory enabled Mme. Curie to attempt trials, such as the Ramsay experiment, that Rutherford and other researchers had declined for want of sufficient radium. She already had another project in mind, regarding the genealogy of the radioelements.

After an atom of radium gave up an alpha particle and changed itself into an atom of gaseous emanation, the emanation gave up an alpha to become a radioactive solid provisionally named "radium A," which in turn transformed by alpha emission into "radium B." Over the ensuing days, weeks, months, and years, further transmutation occurred via the expulsion of alpha or beta particles, resulting more or less quickly in the creation of "radium C," "radium D," and "radium E," then polonium, and lastly the stable end product, lead.

The rightful places of radium and polonium on the periodic table, which Marie had defended repeatedly, differed from the static positions of silver, say, or gold. Instead of a permanent stop, each radio-element's slot represented a waypoint on a long road from one place to someplace else.

In the same manner that radium gave rise *to* other elements, it apparently arose *from* other elements, and ultimately from uranium. The fact that radium never occurred alone in nature, but always and only in the presence of uranium, suggested their ancestral connection. Indeed, a few researchers in the United Kingdom and the United States were actively trying to "grow" radium from uranium in their labs. They ascribed their failures so far to the presumably Methuselah-like half-life of uranium, which was believed to extend over eons. The same researchers further suspected that some number of intermediate elements stood between uranium and radium in the line of descent.

Another route to unraveling the genealogy of radium lay in the variety of uranium ores at widely separated locations around the world. If the same ratio of radium to uranium prevailed in every case, then the common proportionality would demonstrate the mother-daughter—or mother-granddaughter—relationship.

As soon as Ellen Gleditsch returned to Paris from her 1908 summer vacation with family in Norway, Marie pressed her to take up the question of the radium–uranium ratio.

Chapter Nine

LUCIE (Helium)

————

ELLEN GLEDITSCH had been born in Mandal, a small town at the southern tip of Norway, in a wooden house near a beach on the North Sea, and grown up in Tromsø, an island village a thousand miles to the north, where her father served as headmaster of the local secondary school. Despite a childhood spent in remote, bucolic settings, she loved returning to the vigor and energy of Paris, to the city's extravagant spread on both sides of the Seine. She lodged at a pension on the rue Berthollet, one of many Parisian streets named for scientists, about equidistant between the Curie lab and the Sorbonne. Frugal by upbringing and necessity, she covered her living expenses with the several stipends she had won, including a Queen Dowager Joséphine's Scholarship of four hundred Norwegian kroner. And because Ellen's pharmaceutical acumen fitted her to perform the fractional crystallization procedures essential to the laboratory, Mme. Curie did not charge her the usual fees paid by other students and independent researchers.

Back at work in Paris in the autumn of 1908, Ellen was no longer the only twenty-something woman in the Curie lab. Lucie Blanquies, one of Madame's first physics students at the *École Normale Supérieure de jeunes filles* in Sèvres, had joined as a *travailleur libre*. Now a professor of physics at a female teaching academy in Versailles, Mlle. Blanquies felt the need for more guided laboratory experience of the kind Mme. Curie could provide. A part-time return to research, she hoped, would revive her experimental skills, improve her own pedagogy, and enliven the handbook she was revising for advanced physics classes. Like Ellen Gleditsch, Lucie Blanquies enjoyed translating scientific theory and terminology into language that even a beginning student could understand.

"Physics forms a part of the study of natural phenomena," Lucie wrote in the introduction to her handbook, "that is to say, of things that happen all around us. A stone falling, a mirror reflecting our image, water coming to a boil, are some examples of familiar things we must examine. But the study of physics will not merely explain the ordinary events in our everyday encounters; it will conduct us into a new world, utterly unsuspected by the uninitiated." In living proof of that promise, she found herself in the rue Cuvier, comparing the alpha emission of polonium with that of radium C and actinium B. Although all alpha particles were presumed equal, they emerged from different materials at different speeds and traveled different distances, thus ionizing the surrounding air to different degrees.

Ellen, whose first radioactivity challenge had concerned work done at Sir William Ramsay's lab in London, turned now to the radium-uranium studies by American radiochemist Bertram Boltwood of Yale University. She was learning how, as she put it, "a problem emerges in one laboratory, is taken up by a second, and maybe finds a solution in a third." This shared process, at once cooperative and competitive, typified for her "the rivalry to which the progress of science is bound."

In the science of radioactivity, the scarcity of radioelements quickened the sense of rivalry, making Mme. Curie the envy of her peers. Other researchers coveted her mother lode of radium. Boltwood, for example, in a 1904 letter to Ernest Rutherford, reported rumors of "wild scientific orgies" at the Curie lab, featuring obscene amounts of radium bromide. In 1908 Boltwood jealously eyed radium stores in Austria that "made my mouth water," since he had to make do with "insignificantly small amounts" and even "homeopathic doses."

The raw uranium ores that Boltwood sought for his assessment of the radium–uranium ratio were likewise hard to come by. As he pointed out in the *American Journal of Science*, "uranium minerals are extremely rare," so that "the obtaining of suitable samples of the various specimens and varieties is either extremely difficult or altogether impossible."

Boltwood based his first determination of the global abundance of radium vis-à-vis that of uranium on just eight borrowed samples of ores. Seven of these came from the United States—Colorado, Connecticut, and North Carolina—and the eighth from Saxony. His analysis showed they all contained about the same tiny proportion of radium to uranium. Encouraged, he repeated the effort with a dozen additional samples, including some from Norway and Brazil, and still the proportion held constant. In collaboration with Rutherford, Boltwood calculated the worldwide radium-uranium ratio to be 0.00000034. This extremely low number carried great weight because it offered proof of the elements' familial relationship. Whether the one had engendered the other directly or a generation intervened between them, they were kin.

To confirm or contradict Boltwood's conclusions, Ellen Gleditsch reviewed the assumptions he had made and the procedures he had followed. She agreed that the radium content of an ore could best be assessed by pulverizing some of the rock and dissolving the powder in acid, then collecting and measuring the gaseous emanation it released. Since radium emanation was known to be the daughter of radium, it could be used as a gauge of radium content. Some emanation would be lost in the collection process, however, and she thought Boltwood might have over- or underestimated those amounts. She further questioned whether he had allowed enough time and executed all the steps necessary to ensure the emanation's ideal retrieval.

Thanks to the longstanding relationship between the Curie lab and the *Sels de Radium* factory, Ellen had access to an array of ores from far-flung locations explored by prospectors for Émile Armet de Lisle. She chose three types with which to begin her series of analyses: autunite from France, thorianite from Ceylon, and pitchblende from the St. Joachimsthal mine in Bohemia—the source of the Curies' original polonium and radium.

MME. CURIE TOOK a personal as well as a scientific interest in every aspect of polonium and radium research. She was engaged in two separate investigations—one seeking the spectrum of polonium, the other regarding the half-life of radium emanation—when word reached her of a new "radium institute" to be built in Vienna. Her vision of a larger laboratory instantly transformed into a comparable research institution. Such an establishment belonged in Paris, the city where radium had been discovered, named, characterized, and weighed. Her imagined "Curie Institute" would continue its affiliation with the Sorbonne, of course, and perhaps attract the backing of the national government. At the helm of such an organization, she could realize all her ambitions for Pierre and for herself. Having shed much of the self-doubt that taunted her when she first stepped into Pierre's place, she now felt uniquely qualified to fill the director's role.

At the same time, however, the fact of her womanhood threatened her further advancement. Given the popular perceptions of women's place in society, it was easy for people to see Marie as merely Pierre's wife—a dutiful assistant who had made no independent discoveries either before or after his death. There was evidence to support this erroneous view. In June 1903, for example, when the couple visited London, only Pierre spoke at the Royal Institution. In fact, Marie was barred from lecturing in the venue, as all women were. And so, although Pierre, as always, had stressed Marie's specific contributions to their joint work, his spokesmanship translated readily into sole authority. Months later, Pierre unwittingly compounded that impression by returning to London alone to accept the Davy Medal awarded to the Curies by the Royal Society. His solo arrival implied that only he had won the medal, when in truth Marie stayed home because she was ill with the grippe and still recuperating from her miscarriage. The Curies' receipt of the Nobel Prize at year's end, calling the world's attention to Marie as a woman scientist, failed to relieve her of her mistaken identity as Pierre's subordinate, Pierre's helper.

While in England for the event at the Royal Institution, Marie and Pierre met Hertha and William Ayrton—another pair of physicists married to each other. As Hertha explained in numerous interviews over the years, she and her husband intentionally pursued separate avenues of research in order to save Hertha from the fate of being presumed her husband's lackey. Thanks in part to their division of labor, Hertha became, in 1899, the first female member of the Institution of Electrical Engineers—a professional brotherhood uniting more than 3,300 men—and, in 1906, the first woman awarded the Royal Society's golden Hughes Medal for her original investigations of electric arcs and patterns of fluid motion.

In March 1909, when the British press repeated the familiar trope that the lion's share of Curie glory belonged to Pierre alone, the recently widowed Hertha rushed to Marie's public defense. "Errors are notoriously hard to kill," she conceded in a letter to the editor of the *Westminster Gazette*, "but an error that ascribes to a man what was actually the work of a woman has more lives than a cat." The newspaper acknowledged its gaffe two days later, thanking Mrs. Ayrton and citing Mme. Curie as the true discoverer of radium, but said so halfway down a column of minor news items set in small type.

Within the Curie lab, Marie's authority went largely, though not entirely, unchallenged. She judged only one staff member, Jacques Danne, to be bent on his own agenda, at cross purposes with hers. Danne had begun his scientific career as Pierre's student, progressed to the post of *préparateur* in the industrial school, and then moved with the Curies to the rue Cuvier. He had also served as editor of *Le Radium* since its inception in 1904. Although nominally Madame's full-time assistant, he discharged sundry commercial duties for Émile Armet de Lisle and had lately set up a separate radioactivity laboratory for himself. Thus far she had swept his habitual absences under the rug of their long association. She had even done Danne the favor of taking on his younger brother, Gaston, as an independent researcher, teaching him her techniques and giving him valuable experience in handling radioelements. But neither her tolerance nor her generosity had served to remind Jacques of his obligations. Rather than dismiss him, she asked the university administration to provide an additional *préparateur.*

The Faculty of Science turned down Mme. Curie's request on the grounds that Jacques Danne already fulfilled that role. "Having reflected on your actual situation," she was compelled to tell him on June 18, 1909, "as well as on the needs of the laboratory, I judge that you are no longer able to render me the services that I deem indispensable, and I beg you to tender your resignation as *préparateur* immediately."

Danne complied without argument. His official departure from the Curie lab gave him permission and incentive to expand his own laboratory facility. Marie was doubly pleased. Danne's resignation freed her to hire someone new to perform the required duties. Moreover, she recognized the need for a wider pool of radioactivists to meet the demands of research, industry, and especially medicine. Going forward, she anticipated her contact with Danne would be pleasantly limited to discussions of articles for publication in *Le Radium*.

An abundance of such articles emerged from the year's activity in the Curie lab. They described Charles Lattes's design of a new measuring instrument, Bela Szilard's latest findings about radio-lead, and André Debierne's ongoing work with radium emanation, as well as investigations by William Duane, Louis Frischauer, Miroslaw Kernbaum, and Léon Kolowrat, plus progress reports from Ellen Gleditsch and Lucie Blanquies.

———

IN THE COURSE of subjecting her initial ore samples to an elaborate analysis protocol, Ellen realized that her projected study of some two dozen mineral ores, carried out with her usual care, would take years, not months, to accomplish. For now she hoped to publish her preliminary findings on the first three types of ore.

Mme. Curie could practically guarantee publication in *Le Radium* to her assistants and associates, given her frequent contributions to its pages and her position on its editorial board. She had no such inside access, however, to the *Comptes rendus* of the *Académie des Sciences*. No one at the *Laboratoire Curie* save Pierre had ever gained *Académicien* status, and therefore none was entitled to attend the weekly meetings of the *Académie* or to submit notes regarding research results.

Marie's enthusiastic early supporter and co-Nobelist, Henri Becquerel, had died the previous summer, so she turned again to Gabriel Lippmann, her original champion at the Sorbonne. As before, Lippmann obliged her by presenting the reports from her laboratory to the members of the *Académie*. He commanded even greater prestige among his fellows these days as France's newest Nobel laureate in physics, cited for his color photography process. The 1908 Nobel Prize in Chemistry had gone to Ernest Rutherford, in recognition of his insight into the radioactive transformation of elements. As a physics professor, Rutherford was amused by his own sudden "transformation from physicist to chemist" at the hands of the Nobel Committee.

The publication of Lucie Blanquies's research in the *Comptes rendus* and *Le Radium* coincided with the publication of her physics handbook, *Traité de Physique* (second edition). As she explained in the foreword, the text covered the whole curriculum approved for third-year physics courses and provided new supplemental material to help aspiring teachers pass the examination for an advanced degree. There were now enough young women in this category, she said, to justify the inclusion. She had taken the further liberty of using large type to emphasize the sections she considered the most important—or, in some cases, the most meaningful to her. She had relegated the more abstruse concepts to small print as a way to signal to beginners that they could skip over those parts for the time being. An author's note on the cover identified Mlle. L. Blanquies as a graduate of the Sèvres academy, a holder of the advanced teaching certificate in sciences, and professor at the *Collège de jeunes filles de Saint-Germain-en-Laye*.

Mme. Curie's teaching activity in the little cooperative school venture came to an end in June 1909. Irène, the school's raison d'être since 1907, was ready to take on more rigorous preparation for university. A few other adolescents in the group had likewise outgrown the school, and Ève, not yet five, was too young to enter it, so Marie disbanded it. She enrolled eleven-year-old Irène in the *Collège Sévigné*, a private girls' school that best aligned with her personal educational philosophy of allowing students ample free time and lots of physical exercise.

The instruction that Ève required was of an altogether different kind. She did not share her sister's scientific curiosity, but instead showed an innate musical ability. At age three she had toddled over to the piano and picked out the tune of "*Au clair de la lune.*" By four she could play back, or sing, any melody she heard, and had memorized about thirty popular songs. Mme. Chavannes, who had taught English, German, and geography in the cooperative school, now came to the Curie home to give Ève voice lessons. Marie considered the child a prodigy.

"Like a great many children," the adult Ève mused in her biography of her mother, "we were probably selfish and inattentive to shades of feeling. Just the same we perceived the charm, the restrained tenderness and the hidden grace of her we called—in the first line of our letters spotted with ink, stupid little letters which, tied up with confectioners' ribbons, Marie kept until her death—'Darling Mé,' 'My sweet darling,' 'My sweet,' or else, most often, 'Sweet Mé.'"

Marie's sister Helena, now a school headmistress in Warsaw and married to photographer Stanislaus Szalay, arrived in Paris in July 1909 with her eleven-year-old daughter, Hanna. They scooped up Irène and Ève and traveled on to Saint-Palais-sur-Mer, near Royan on the southwest coast. "One day, Hania leads us on walks where she likes, and the next day it's my turn to say where we go," Irène wrote home to "sweet Mé." On the third day, she reported, the choice of walking route fell to Mlle. Edwige, the Polish governess, and then to Aunt Helena the day after that. "In this fashion, each one of us gets her way once every four days."

Marie's letters to Irène as often as not contained math problems, in continuation of the teacher-student side of their mother-daughter bond. Irène responded to one of these challenges by admitting, "I've forgotten a little what you have to do to get the derivation of a radical and of the two numbers that divide it. You would do well to send me the rules of derivations and some examples." From the seaside, she also implored "my dear Mé" for reports of her pet mice, Filou and Tigrette, and the condition of certain favorite trees in the garden.

Marie made only a brief visit to her children at the seashore that summer. Lab duties consumed her time, and her father-in-law fell ill with pneumonia. She tended to him lovingly, but he was a doctor by profession and that made him a difficult patient. Irène, who had been the old man's joy since the day she was born, fretted and feared she would lose him, just as she had lost her father.

"What are the flowers on Pé's grave?" she asked toward the end of another letter from Saint-Palais-sur-Mer, and "which ones are in bloom?"

Chapter Ten

SYBIL (Thorium)

MARIE HAD SERVANTS to help her at home now—a Polish governess for the girls and a housekeeper who cooked the family's meals—but she nursed her ailing father-in-law herself. "The year 1909 is the year of the illness of Grandpère," she wrote in a notebook, allowing the simple fact its full weight. Through the summer and early autumn, she spent as much time as she could by his side, "listening to his remembrances of passed years." Nearness to him and caring for him might have offset, to some small degree, the distance that had separated her from her own father in his final days.

At the lab she finessed another small expansion to accommodate a few incoming researchers. An additional space in the Sorbonne Annex was now completely refitted according to her specifications. Only the sign on the door betrayed the room's recent past as storage for *"Collections Botaniques."* The new chemist from England, twenty-two-year-old May Sybil Leslie, noted the labeling mismatch and mentioned it in her first letter home.

Sybil, as she preferred to be called, had never been so far away from home. Born and raised in the Yorkshire village of Woodlesford, she became the first local girl from a working-class family to win a university scholarship. The windfall carried her to Leeds, six miles away, where she took first-class honors in chemistry at her graduation in 1908. She then completed her master's degree and received an "Exhibition of 1851 Scholarship," awarded only to the most exceptional young scientists and engineers. Suddenly she was free to go anywhere. She had zero experience with radioactivity, but an earnest letter of endorsement from Arthur Smithells, head of the chemistry department at the University of Leeds, eased her way into the Curie lab, her chosen destination.

Sybil Leslie found lodgings in Paris at an address that Mme. Curie suggested, then spent the first few weeks acclimating herself to the lab's unfamiliar apparatus, such as the piezoelectric quartz, under the guidance of André Debierne.

"Monsieur Debierne, the discoverer of actinium," Sybil reported to Professor Smithells, "is the *Chef du Laboratoire* and has most of the burden of the students. He is a Frenchman of the most charming type, gentle, kind, and courteous in manner, and with a vast fund of patience which he certainly needs, for the advice offered in every difficulty is '*Demandez à M. Debierne.*'" As a further testament to the man's forbearance, Sybil noted that "he listens very patiently while I murder his language and tie myself in linguistic knots."

M. Debierne and Mme. Curie had recently acquired several more tons of uranium ore residue, now undergoing a new separation process at the *Sels du Radium* factory. They hoped to isolate enough polonium for a conclusive visualization of the element's spectrum.

"There are only two ladies besides myself," Sybil's letter continued, "Norwegian Mlle. Gleditsch, and French Mlle. Blanquies. Of the French lady I see very little because she does not spend all her time here, but of Mlle. Gleditsch I see much since she lives in the same *pension*. She has been exceedingly good to me and has prevented me from feeling lonely."

Her orientation complete, Sybil received a specimen of the mineral thorite, a primary source of thorium, and was asked to discern what other radioelements, if any, the rock might contain. Stationed at one of the new white-tiled lab benches in the added-on room, she performed the necessary chemical analyses with the practiced ease of past experience. But the current challenge required her to test each breakdown compound product for radioactivity. Entirely new entities,

undetectable except by electrometer, might emerge at any
stage in her dissolutions. Perhaps the "mother" or "daughters" of thorium would reveal
themselves.

In the evenings, Sybil
returned to her lodgings in
the rue Berthollet along a
route illuminated by Paris's
emblematic streetlamps. Even
as electric lighting proliferated
through the city, the old gas-
lights still glowed, their light
intensified by the presence of
thorium in their mantles. The
rare metal had been put to the

May Sybil Leslie

purpose of amplifying gaslight decades before Mme. Curie discovered
its radioactivity.

"Mme. Curie spends a considerable amount of time working in the
laboratory," Sybil observed, specifically in the *Salle à Chimie* in the
courtyard, detached from the main building. "She does not appear
to come around much to the students but receives them very kindly
when they seek her. She does not speak English at all, nor does she
appear to understand spoken English except a few scientific terms.
She speaks very quickly and to the point and is very quiet in manner
but by no means languid. She has a face of great intelligence and
the expression is rather worn and sad in general, but she has a most
charming smile at times which quite transfigures her."

Sybil naturally enrolled in Madame's course on radioactivity. By
the end of November 1909, after several weeks of instruction, she
was still finding the lectures "by no means easy to follow." The first
two or three, she said, had drawn numerous curiosity seekers eager
for a glimpse of the famous lady professor. Attendance had shrunk
appreciably since then. "The increasing difficulty of the lectures,"

Sybil explained, "on account of the mathematics introduced and the great speed at which Mme. Curie proceeds, has gradually weeded out the audience, and for the most part only the enthusiasts who are really interested in the subject now come."

———

IN DECEMBER MARIE moved Grandpère's bed—from the bright clutter of his room, where shelves sagged under mementos, medical texts, and the novels of Victor Hugo—into the living room. She set up a cot for herself in the dining room, where she would be sure to hear him should he call out to her during the night.

At this same dark time, she saw the first glimmer of her grand new laboratory. Officials at the Sorbonne contracted with doctors at the *Institut Pasteur* to plan a comprehensive center for radioactivity research. One half would be devoted to the medical uses of radium, the other half to the physics and chemistry of radioelements. Marie would direct this arm of the enterprise, which was to occupy its own separate, several-story building, and be known as *le Pavillon Curie*.

The skin lesions reported early on by the Curies and Becquerel had piqued the interest of the medical community, and that interest had grown along with radium's fame. In 1904 Dr. Antoine Béclère, one of the first physicians to perceive radium's promise for medicine as a destroyer of cancerous growths, brought two long-suffering women into the Curie lab and treated their recurrent breast cancers by direct exposure to the element. By 1909 radiation therapy, also known as "curietherapy," was administered in the hospital, via small glass ampoules of radium emanation laid on the skin surface. In the case of deep tumors, the tiny tubes or needles of radioactive gas could be surgically implanted.

The half-value of radium emanation, now a crucial constraint in medicine as well as a fundamental parameter in physics, had been measured repeatedly over the years. Pierre had determined 3.99 days as the time required for any quantity of the material to diminish by half. Ernest Rutherford later shortened that figure to 3.71 days, and recently researchers in America and Europe had recorded intermediate values of 3.88, 3.86, and 3.75. Marie considered all the factors that might have driven the various results up or down, such as the volume of gas each investigator had studied, the type of device used to quantify radioactivity, the temperature and humidity of the ambient air, and so on. Finding fault with several of the methodologies, she decisively reset the half-value, or "half-life," of emanation at 3.86 days.*

She turned next, early in 1910, to the isolation of radium in its pure metallic state—something never seen in nature or in any laboratory. What would the element look like, forcibly stripped of its chloride or bromide companion, as naked and alone as a disembodied soul?

The year before Pierre died, Marie had accompanied him to a series of séances, where the dead were summoned to speak or otherwise reveal themselves. Such sessions were very popular at the time, and it was typical of Pierre to keep an open mind toward any novel phenomenon, even the trances of a spiritualist. As the couple sat with other participants in the semidarkness around the levitating table, vague apparitions hovered overhead. Marie had seen these things with her own eyes, but also with a sense of how easily the eye could be deceived.

The isolation process that she and Debierne adapted for separating radium from radium chloride was fraught with danger—not merely of failure, but of losing their radium along the way. They began with a pinch of salt, or rather a decigram of radium chloride, which they dissolved in solution so as to separate the elements by electrolysis: They dipped two metal strips, or electrodes, into the solution, one made of mercury, the other of platinum-iridium, and both connected to a battery. Current flowing between the electrodes invited the radium ions to collect on—and combine with—the mercury electrode.

They then sealed the radium-mercury amalgam in an iron capsule, which they heated, very gradually, to boil off the mercury. At 700 degrees, when they feared the radium itself might vaporize, they halted the procedure.

---

*The currently accepted figure is 3.825.

Having at last trapped their quarry, they saw the unalloyed metallic radium shining with a brilliant whiteness. It adhered so tightly to the iron capsule that they had to pry it out with a chisel-tipped tool. A fragment that fell on a sheet of paper burned right through, leaving a black-rimmed hole. Exposed to air, the bright radium darkened in combination with nitrogen. It reacted violently with water, forming an oxide and a blackish residue. When assessed for radioactivity, it behaved as radium always behaved, releasing alpha particles and radium emanation at a rate as regular and recognizable as a heartbeat.

Ever-practical Marie would not consider preserving the pure radium in this bizarre state, when every shred was needed for other purposes. A unicorn could not do the work of a horse. She and Debierne set about turning their hard-won sample back into a tiny pillar of salt.

THE TYPICAL RAIN and chill of the Paris winter grew into a deluge in late January 1910, when the Seine overflowed its banks and flooded some fourteen thousand buildings throughout the city. The main thoroughfares turned into river tributaries. A metre of water covered the tracks at the Gare d'Orsay terminal. Resilient Parisians made the best of things at first, navigating the inundated streets in boats or on the raised wooden walkways hastily built to meet the emergency. After several weeks, however, with no relief in sight, hordes of residents evacuated.

The *pension* where Ellen Gleditsch and Sybil Leslie lived stayed dry, but flooding at the Sorbonne Annex cut off the Curie lab's heat and electricity. Sybil described the scene to Smithells—how only a few of the rooms could be lighted by gas jets, leaving unlit areas to be passed through "candle in hand." The cold made it difficult to remain any length of time on task, and dampness rendered the electroscopes useless. The one Sybil used, she quipped, had become "a veritable hygrometer" (an instrument for measuring the content of water vapor in the atmosphere).

The Curie home in Sceaux lay well beyond the flood zone. Dr. Eugène Curie died peacefully there on February 25. When the family gathered at the local cemetery, Marie prevailed on the gravediggers to bring up Pierre's coffin before burying Grandpère. She wanted the beloved patriarch to be interred nearer to his wife and, by the same switch, to reserve the space just above Pierre for herself.

Irène, who had not witnessed her father's original burial, slumped under the weight of both losses. "She is very shaken," Marie wrote in her journal. "Her sadness is profound. . . . She suffers and matures." Irène's maturity was further hastened by her classmates at the *collège* she now attended in Paris, where her intellectual ability had placed her among students beyond her twelve years. Ève, age five, was enrolled at an elementary school, while also receiving private tutorials from her mother, first thing each morning, and continuing her music lessons.

In March, when the floodwaters receded, Sybil Leslie professed herself astonished to see how rapidly Paris could shrug off its sorrows and "reassume its smiles." With light, heat, and dry equipment, she continued dividing and subdividing her fifty grams of thorite until she arrived at a few tantalizing active traces. These, unfortunately, were too scanty to identify. There was nothing for it but to start over, this time with much, much more material. "After working with grams in beakers & evaporating dishes," she noted, "it is rather difficult to adjust oneself to kilograms in huge jars & earthenware bowls."

In spite of these difficulties, Sybil remained confident and focused on her experiments. Her letters to Smithells bore none of the insecurity or self-doubt that Harriet Brooks had often expressed to Rutherford. In the three years since leaving the Curie lab, Harriet had convinced herself that women in general were ill-suited to scientific research. She aired this belief before the McGill Alumnae Society in April 1910, when its executive committee invited Mrs. Frank Pitcher to speak about her experience of working under Mme. Curie.

"Perhaps it was a feeling of delicacy that prompted the committee to confine my attention to Mme. Curie this p.m., that I might not be forced to expose the poverty of women's contributions to physical science." Harriet did not rue the lack of opportunities for women, but rather blamed the dearth of women scientists on certain shortcomings of her sex. "The combination of the ability to think in mathematical formulas and to manipulate skillfully the whimsical instruments of a physical laboratory—a combination necessary to attain eminence in physics—is apparently one seldom met with in women," she said. She seemed to have forgotten that she herself had once embodied these abilities. Now she took pains to assure her fellow alumnae that even a very large man was far more likely than the average woman to master the delicate and intricate tasks of laboratory science. "It is no rare thing to see a man with hands twice the size of one's own and every appearance of being very dangerous to fragile furniture"—this was Rutherford to a T—"who will handle quartz fibers, so fine that they can be seen only against a black cloth, with the same ease with which we would disentangle a skein of wool."

Even if gifted with greater "deftness of hand," Harriet thought women would still feel indifferent toward the physical sciences due to "their want of human interest." Nevertheless, she said, the science of radioactivity had seen astounding developments in which "Mme. Curie has played an important and unforgettable part." Here she proceeded to describe such arcana as atomic weight determinations, the theory of radioactive transformation, the difference between alpha and beta radiation, and the heat release of radium.

"It is naturally difficult to separate the work of Mme. Curie from that of her husband during the years in which they worked together, but I have always had the impression that she was the pioneer in planning and he the skilled experimenter," Harriet concluded, as though to prove her argument. "At his death his wife was appointed his successor at the Sorbonne, an honor to a woman quite without precedent at that ancient university. The duties of the position, the growth of the research laboratory under her care and no doubt also the loss of her skillful collaborator have somewhat lessened her output of work in the last few years."

In fact, Marie had rebounded after the fallow period immediately following Pierre's death. Actively re-engaged in research in 1910, she was also preparing her textbook on radioactivity for imminent publication, arranging for the addition of another four researchers to her lab, and playing a central role in planning the new *Institut du Radium*. Friends remarked on her rejuvenated appearance. One evening she startled them all by showing up at a social gathering in a white dress instead of her customary black.

Marie was experiencing a renascence of romantic feelings that she thought had died with Pierre. These centered on Paul Langevin, who had taken to confiding in her the details of his unhappy marriage. Marie and Paul had long been friends as well as colleagues at Sèvres, but now, listening to the litany of his distress aroused her sympathy and awakened her desire. Although Paul's wife stood between them, it seemed that Pierre bound them the more tightly together through their shared memories of him in happier times.

In mid-July, after Irène and Ève left for a beach vacation in Brittany with their aunt Helena, Marie visited Paul in the two-room apartment he rented on the rue Banquier. There, as discreetly as possible, they became secret lovers.

"My dear Paul," she wrote to him later that summer from the Brittany coast, "I spent yesterday evening and night thinking of you, of the hours that we have passed together and of which I have kept a delicious memory. I still see your good and tender eyes, your charming smile, and I think only of the moment when I will find again all the sweetness of your presence."

Chapter Eleven

EVA (Radium)

———

DISCRETION, IT TURNED out, proved no guarantee of secrecy. Jeanne Langevin grew suspicious of her husband and looked for evidence of his infidelity. When she intercepted one of his notes to Marie, toward the end of August 1910, she berated him for his betrayal. Then she accosted Marie on the street, threatened her with violence, and demanded that she leave the country.

Too upset to return alone to Sceaux after this distressing encounter, Marie ran to the home of her friends Henriette and Jean Perrin on the boulevard Kellerman. The Perrins, who had gathered with the Langevins and the Curies on so many Sunday afternoons, agreed to intercede as trusted mediators, to restore peace and avoid scandal.

Marie stepped back. It was clear to her that Paul's marriage jeopardized his scientific genius, whereas she valued and hoped to nurture it—after he left the marriage and formed a new life with her. Meanwhile she would help him find the strength, the words, the specific actions he must take to extricate himself.

Although the lovers swore to stop seeing each other in private, they could hardly avoid social contact, given their professional ties. In mid-September, Paul and Marie, along with Jean Perrin, André Debierne, and four young researchers from the Curie lab, traveled to Brussels for the second International Congress on Radiology and Electricity. The first such congress had been held in 1905 in Liège. The fact that a second, much larger one followed in 1910 gave testimony to the rapid progress in the field and its influence on the progress of science. Here Marie faced admirers and rivals of a different stripe. The event drew five hundred physicists, chemists, and medical doctors from all parts of the world. These eminent men of science included every radioactivist from Ernest Rutherford and Bertram Boltwood to Stefan Meyer of the new Viennese Radium Institute, which was due to open in just a few weeks.

On the first afternoon of the congress, the organizers invited Mme. Curie to recount her recent success in isolating pure metallic radium. The discomfort with public speaking that came over her at the podium subdued the tone of her delivery, but she nevertheless impressed the audience by stating the simple facts of the task and the risks it entailed. In less skilled hands, she led her listeners to understand, the precious substance could have been lost.

Later that day she met with nine other leaders in radioactivity research to plan the creation of an international radium standard. This much-desired touchstone would anchor qualitative and quantitative measurements of the preeminent radioelement, whether for research, for sale, or for curing disease. At present, each laboratory had only its own standard to rely on, and this situation prevented the meaningful exchange of data. Rutherford, who had gone about testing the in-house standards at several institutions, found alarming differences among them. The times demanded a single sample of pristine radium salt, sealed in a suitable container, to serve as the paragon for all comparisons, much the way the international metre bar, held at the Bureau of Weights and Measures in Paris, justified all length determinations.

Paul Langevin

Marie had a cold. She felt achy and dull-headed, but also

adamant that the honor—and onus—of fashioning the international radium standard belonged to her. The others concurred, or acquiesced. The effort would likely occupy a year of her time, she estimated, to gather and purify the necessary amount of radium chloride, at considerable expense. Rutherford, as chairman of the select committee, assured her she would receive compensation from professional societies and the governments of countries that were stakeholders in the young science.

Some argument arose as to the ideal mass of the standard. Marie proposed twenty milligrams—an amount that struck some in the group as excessive. She held firm: nothing less than twenty milligrams would do, given the difficulty of weighing small quantities of radium salts with the requisite accuracy. Here again, she won consent. In a third development in her favor, the committee moved to define a new unit of radioactivity and name it the "curie" in honor of Pierre.

Marie stood on the platform during the final full assembly on Thursday, when the resolution regarding the radium standard was read aloud in French, German, and English. That evening she attended the banquet arranged for the congress participants in the *Hôtel de Ville*, followed by a performance at the national opera house, the *Théâtre Royale de la Monnaie*, where she sat next to Rutherford. At intermission, he noticed how unwell she looked and insisted on escorting her back to her hotel room to rest. "I think she has been overworking," Rutherford opined later to Boltwood in an informal postmortem, "and some of the medical fraternity consider she is in a very bad nervous state."

On Friday morning Marie took the train to Paris, then continued on to Brittany. Irène and Ève awaited her on the rocky beach at l'Arcouest, along with her sister Helena and niece Hanna. The last days of summer were still warm enough for swimming, and the evenings held a pleasant tumult of adults and children in the rented cottage that Marie's brood shared with the Perrins.

She wrote to Paul several times at the rue Banquier address. "It would be so good to gain the freedom to see each other as much as our various occupations permit, to work together, to walk or to travel together, when conditions lend themselves. There are very deep affinities between us which need only a favorable life situation to develop."

After the vacation ended and the families returned to Paris, all efforts to keep Mme. Langevin calm fell to pieces. Jean Perrin was forced to inform his friend Paul that "your wife has recently made threats of murder so serious as to convince us that she is capable of acting on them." He and Henriette would welcome Paul into their home at any time, he said in his note, but Jeanne had made herself a pariah by her "incessant threats of murder." These outbursts had frightened and preoccupied the Perrins to the point where "it would be laughable to speak any more of friendship."

Marie carried on with her work as best she could in the tense atmosphere, not knowing how Jeanne would act or what Paul would decide. As for herself, though she had often been dismissed in the past as simply her husband's assistant, she now stood more firmly on her own merits. Soon construction would begin on the nascent French radium institute, at a site selected between the rue d'Ulm and the rue Saint-Jacques in the fifth *arrondissement*. The population of the Curie lab had swelled to twenty-two, or about twice the number in 1906 when she took over as director.

Among the new recruits, Eva Ramstedt of Stockholm stood out. At thirty-one, she was older than the typical newcomer and already held a doctoral degree from the University of Uppsala. Her father served in the Swedish cabinet, including a brief term as prime minister during the dissolution of Sweden's union with Norway. Eva Ramstedt's family politics, however, did not keep her from forging an immediate bond of Scandinavian fellowship with Ellen Gleditsch. Through Ellen, she developed a friendship with Sybil Leslie as well. Their former coworker Lucie Blanquies had parlayed her Curie-lab experience into a promotion at the school where she taught, the *Collège de jeunes filles de Saint-Germain-en-Laye*. Harriet Brooks, now Pitcher, gave birth to a daughter, Barbara Anne, on October 19, 1910.

That fall Marie made the final adjustments to the manuscript of her book, *Traité de Radioactivité* by Madame P. Curie, and saw it published in two volumes by Gauthier-Villars, a specialist in scientific texts. Not only did she identify herself as "Madame Pierre Curie" in her byline, as opposed to "Marie Sklodowska Curie," she also inserted a photograph of Pierre as the frontispiece, instead of her own likeness. It was the same picture she had chosen for the frontispiece of his complete works, *Oeuvres de Pierre Curie*. She remembered the day the photo was taken, just weeks before his death. The studio photographer caught him looking directly into the camera, resting his chin on his hand and wearing a placid, contemplative expression—the very portrait of a dreamer.

Marie's recollections of Pierre, her adherence to his way of thinking, still guided many of her actions. In November, for example, when the National Order of the Legion of Honor offered her the chevalier cross, she rejected the recognition in deference to her husband. She well remembered Pierre's distaste for "decoration"—how he had used his refusal of Legion laurels as an opportunity to request a laboratory. Marie spoke only of Pierre. She could not accept a prize that he had spurned.

No sooner had she declined the Legion of Honor than she was courted as a candidate for election to the *Académie des Sciences*. The death of *Académicien* Désiré Gernez had created a vacancy in the physics section, and Mme. Curie's colleagues urged her to contend for it. Membership in the *Académie* conferred a distinct national honor—not an empty title but a coveted position of strength. As a member she would be allowed to present the work of her laboratory to an audience of her peers at the weekly meetings, and to influence the distribution of important grants and prize money. Those privileges had convinced Pierre to join the *Académie* in 1905. Since he had been a member, her election, should it come about, would not dishonor his memory or sacrifice their shared principles. Her biggest obstacle was her sex. A woman's entry into the exalted ranks of the all-male *Académie* would reverse a tradition of more than two centuries' standing.

Mathematician Gaston Darboux, a permanent secretary of the *Académie*, endorsed Mme. Curie's candidacy in a long letter to the editor of *Le Temps*, published on December 31, 1910. Not only had Mme. Curie shared in the work of her late husband, the letter said, and led their laboratory since his lamented death, but she was the first woman to achieve the standing of a Sorbonne professor and the first woman to gain admission to the Physics Society. Under her leadership, the Curie laboratory had grown populous and prosperous. Moreover she had recently published a book and been singled out to create the international standard of radium. "If anyone were to doubt the long-range promise of the work so well begun," Darboux argued, "it would suffice to look at what has happened abroad. In England, America, Germany, Austria, etc., institutions are being built to pursue the research begun by the Curies." The "prodigious outpouring of esteem" for Mme. Curie had already gained her honorary or foreign-member status in the scientific societies of Stockholm, London, St. Petersburg, Prague, Bologna, Cracow, Philadelphia, and other cities, as well as prizes and doctoral degrees from a dozen foreign universities. "While it often falls to our country to march ahead of other nations, this time it is the foreigner who sets an example for us."

On a recent trip to Rome, Darboux said he had seen archaeologist Ersilia Lovatelli formally admitted to the Lyncean Academy—the Italian equivalent of the French Academy of Sciences. The Berlin Academy of Sciences had also recognized a woman, Elise Wendel, as a scientific equal. "The moment has come," he concluded, for France to follow suit.

Within days, the issue of female candidacy came up for debate at a plenary session of the five learned academies constituting the *Institut de France*. These were the *Académie Française*, guardian of the French language, and the academies of sciences, humanities, fine arts, and moral and political sciences. The controversial discussion topic doubled the usual attendance at such meetings. Some 150 men, attired in their embroidered green ceremonial garb, assembled at Mazarin Palace on January 4, 1911. They voted eighty-five to sixty *against* the eligibility of women, citing "immutable tradition." Nevertheless, since the *Institut* had never before violated the autonomy of any component academy, the *Académie des Sciences* felt free to ignore the outcome of the vote on the *candidature feminine*. Thus Marie's name appeared at the top of the list, made public on January 17, of the seven candidates vying for the physics seat.

Her chief competition for *Académie* admission lay with Édouard Branly, an inventor and professor at the *Institut Catholique de Paris*. Branly's achievements did not really compare to hers, though naturally his supporters thought otherwise. They lauded his contribution to wireless telegraphy almost to the exclusion of that technology's more widely cited inventor, Guglielmo Marconi of Italy. Branly, born at Amiens, was a thorough Frenchman, while Marie Sklodowska Curie, a resident of Paris for nearly twenty years now, was still a foreigner who spoke with an accent. Branly had waged two previous campaigns for *Académie* election. His age, sixty-six, hinted this might be his final opportunity, which gave him another sympathetic advantage over Mme. Curie, who was only forty-three.

Pierre had failed in his first attempt to enter the *Académie*, in 1902. Serious and straightforward as he was, he detested the perfunctory social calls that a supplicant was expected to pay on the senior members in order to secure their votes. At each one's doorstep, Pierre had half hoped to find the individual not at home, so he could simply leave a *carte de visite* with a note to show he had tried. At the time of his second foray, in 1905, the *Académie* seemed more eager to acquire him than he was to apply. By then he had won a Nobel Prize and a base of previous backers to lend the necessary support, but even so, custom required him to make the social rounds.

Everyone considered such interviews *de rigueur* for *Académie* admission, but for Mme. Curie to meet one-on-one with men in their private apartments would seem an awkward breach of etiquette. She conducted some number of such visits, and the stalwart Gaston Darboux made several more on her behalf.

Although Darboux's initial public letter had expressed Mme. Curie's hope that her candidacy would avoid commentary in the press, the opposite occurred. She became a lightning rod for public opinion on both sides of the issue. To feminists, Mme. Curie stood as proof that genius had no gender. To traditionalists, she was seen by turns as presumptuous, dangerous, and unwomanly. Her name evoked strident outcry from constituencies far removed from science, including religious factions and social movements mired in political conspiracies. The notorious Dreyfus affair of the mid-1890s, in which a Jewish officer of the French Army had been accused of espionage and falsely convicted of treason—twice—before his ultimate exoneration, still divided the extremes of French society into pro- and anti-Dreyfus camps, prompting the right-wing newspaper *l'Action française* to cast Mme. Curie as a stand-in for Alfred Dreyfus.

On Monday, January 23, the day of the contentious election, nearly every one of the sixty-six members of the *Académie des Sciences* reported to the *Institut* headquarters. Mme. Curie's supporter Charles Bouchard traveled all the way from Cannes to cast his vote for her. He found a heavy cordon of guards at the gate, fending off crowds of spectators. At first only members were let into the building. A little later, the doors were opened to nonmembers as well, but only men gained entry. The exclusion of women gave an early sign that "immutable tradition" would carry the day.

Inside, the first order of business concerned the regular weekly reporting of scientific work, which filled the time till 4:00 p.m., the hour appointed for the vote. Then each member marked a secret ballot, and guards collected the paper slips in large urns. As *Académie* president Armand Gautier read the names aloud, secretaries recorded the tally and everyone else kept a private mental count. Twenty-eight members voted for Mme. Curie, twenty-nine for Branly, and only one for another candidate. A second balloting settled the suspense. This time there were thirty votes in Branly's favor, a clear majority. Since Branly was present, having entered the hall with the crowd, President Gautier invited him to step forward and take his place among the company that now counted him one of their own.

News reached the Curie lab quickly. Ellen, Eva, and Sybil, who had chipped in to buy flowers that morning for presentation at the moment of Madame's victory, enlisted the laboratory mechanic, Louis Ragot, to retrieve the bouquet from its hiding place under the precision balance table and get rid of it. After Marie learned of her defeat, she made no public comment, nor any private one that anyone could recall.

Chapter Twelve

JADWIGA and IRÉN
(Gold)

MOST NEW STUDENTS and researchers joined the Curie lab at the start of the fall semester, but Jadwiga Szmidt arrived in January 1911, amid the excitement and disappointment of the *Académie* vote. She had been teaching at a girls' school in St. Petersburg, having grown up, as Mme. Curie did, in an academically oriented family under Russian domination in what had once been Poland. She was twenty-one years old and capable of discussing physics in Polish, Russian, English, German, French, and Italian.

As Jadwiga Szmidt absorbed the techniques specific to radio-activity, she overheard quiet conversations in the lab concerning the great injustice dealt to Madame by the *Académie des Sciences*. Some workers predicted that "*la patronne*" would gain admission on her next attempt. But the elite position of *Académicien* carried a lifetime tenure, so openings arose only occasionally. No one could say when another opportunity might present itself, or whether Mme. Curie would seize that opportunity. For now, she buried the incident under a weight of other obligations, such as the creation of the radium standard and the construction of the new radium institute under the joint auspices of the Sorbonne and the *Institut Pasteur*.

Marie had dreamt of a laboratory in the countryside, where one could take breaks in the open air. Since the chosen location for the new institute on the rue des Nourrices stood well within the city limits, she began planting rose bushes and linden trees there, even before the first foundation stones were laid. Over the next two or three years, she reckoned, the saplings and rambler roses would grow along with the tan brick-and-stone structures, ready to provide shade and bloom by the time of the official opening. As soon as the buildings began to take shape, Marie visited the site regularly and met with the architect, Henri-Paul Nénot. Her input guided every decision regarding the *Pavillon Curie*, from the large size of the rooms to the height of the tall windows and the inclusion of an elevator, which she deemed a must.

As the uneasy separation from Paul Langevin wore on, she continued to communicate with him by letter. At Easter time their letters to each other, which had accumulated in a drawer at the rue Banquier flat, were stolen. Shortly after the theft, Mme. Langevin's brother-in-law informed Mme. Curie that her compromising letters were now in his possession and would soon be used to ignite a scandal.

Days passed, then weeks, and still the threatened exposure did not occur. Marie enjoyed a pleasant diversion in mid-May when her friend Hertha Ayrton came to Paris to address the Physics Society. Her invited lecture, about the formation of sand ripples on the sea floor, included lively demonstrations with large, oscillating tanks of water, every gallon of which had to be carried up to the fourth-floor amphitheater for the event. Mrs. Ayrton made the rotation of the eddies visible by throwing handfuls of pepper, bronze powder, or drops of paint into the tanks at key moments. On the weekend, Alice Chavannes hosted both Hertha and Marie at a luncheon party in their honor, and then Hertha went home with Marie to visit Irène and Ève.

Ève's musical talent continued to be a source of pride for her mother. In June Marie arranged for the child, now six, to perform for Ignacy Jan Paderewski, the renowned pianist and composer. "Paderewski thinks she has exceptional ability," Marie crowed in her journal, quoting his exact words in Polish. "I suspected as much. Though I understand nothing of music, I felt strongly that she didn't play like just anybody."

Marie had kudos to spare for her laboratory "daughters" as well, especially Ellen Gleditsch, who was still analyzing ores in pursuit of the radium-uranium ratio. Unfortunately, Sybil Leslie had not yet published any of her work with thorium, which meant that her financial support from England—her Exhibition of 1851 Scholarship— would end with the summer instead of extending through a third year in France.

"I think my work will be in a fit form for publication when I leave here," Sybil wrote to her mentor, Arthur Smithells. "It is really suffering from too much radioactivity. A number of people seem to be employing radium emanation at present and my electroscope is disagreeably sensitive to the influence of anyone entering from the 'salle active' so that I spend half my time in keeping dangerous people out & in airing the room. Formerly more care was taken to prevent the distribution of activity all over the laboratory but as the foundations for a new Institute of Radioactivity for Mme. Curie are now laid, all precautions seem to have been abandoned."

Despite her concerns, Sybil managed within a couple of weeks to ascertain the atomic weight of thorium emanation. She succeeded by following the same procedures and using the same apparatus that André Debierne had employed to determine the atomic weight of actinium emanation. She presented her results to the *Académie des Sciences* at the end of July via chemist Paul Villard, a good friend of the Curie lab and famous for the discovery of a third type of radioactivity emission, "gamma rays." With one paper published in the *Comptes rendus*, Sybil submitted another to *Le Radium* on the solid breakdown products of thorium. In light of these achievements, the scholarship committee agreed to support her research for another year, which she arranged to spend back home in England—at Manchester, with Ernest Rutherford.

Marie decamped briefly in July to collaborate with Dutch physicist Heike Kamerlingh Onnes at his cryogenic laboratory in Leiden. Only there could she expand her tests of radium's intrinsic properties into the extremes of frigid temperatures, near absolute zero. When she returned from the Netherlands, she learned of the latest broken link in the Langevin marriage bond: A violent argument had driven Paul to leave the house with his sons, Jean and André. After consulting a lawyer, he took the boys to London for a month, and during that time Mme. Langevin filed suit against him for abandonment.

The prospect of an ugly public trial colored the rest of that summer, including Marie's long-planned reunion with her family in a high, green retreat near the Dluskis' sanitarium. She sent her daughters east with their governess soon after the school semester ended, promising to join them for her own much briefer vacation later. Meanwhile she all but locked herself in the lab, working evenings and Sundays to finish her preparation of the radium standard.

"We eat ice cream twice a week," Irène reported from Zakopane. "When you come here, I hope you will keep Aunt Helena from trying to make me eat, eat, eat because I can't eat so much."

Ève wrote, too: "Sweet Mé! Every day I do my gymnastics morning and evening after I practice my music, then reading and handwriting. I don't have friends my age but I'm going to and I would very much like to know when you will come." While waiting for Mé, the sisters rode horseback for the first time and picked wild berries as their mother had done in her own girlhood. When Mé finally arrived, she led them on hiking and camping excursions in the Tatra Mountains. And when it was time to go home, they took along Cousin Hanna, age fourteen, to spend the year living with them and attending school in Paris.

Fall 1911 brought Ellen Gleditsch back to the city for a fifth year in the Curie lab. She had recently been named a university fellow at her home institution in Norway. The new position came with teaching responsibilities, but Ellen had been exempted from these for the current academic year in order to complete her *licenciée* at the Sorbonne under Mme. Curie. Madame's demonstrated faith in her had confirmed Ellen's commitment to research. Now in her early thirties, Ellen could not imagine dividing her time between her chosen work and any man. When questioned on the subject of marriage, she would simply say that "chemistry is my everything."

The familiar laboratory looked different without her friends and housemates. Eva Ramstedt had returned to Stockholm, Jadwiga Szmidt to St. Petersburg, and Sybil Leslie to England. But Ellen found another newcomer to put at ease, to coach from her own experience, and to learn from: Irén Götz, twenty-two, had just crowned her chemistry studies at the University of Budapest with a brilliant thesis on the quantitative measures of radium emanation.

———

At the end of October Marie saw Paul again for the first time in months. They met in Brussels—not for a secret tryst but as two prominent scientists invited to participate in an exclusive private discussion of the latest developments in physics, hosted by the successful industrial chemist Ernest Solvay. Unlike the previous year's international congress, which had drawn hundreds of practitioners to the Belgian capital, the 1911 Solvay Council grouped its twenty-four attendees around a single conference table in the Hotel Metropole. Marie and Paul were flanked by their friends and countrymen Jean Perrin and Henri Poincaré, as well as by Ernest Rutherford and James Jeans from England, plus a dozen other researchers of equal repute, including the thirty-two-year-old Albert Einstein, a professor of theoretical physics in Prague. As usual, Marie found herself the only woman present.

If the Langevin marriage was in crisis, so too was the science of physics. Its foundations had been shaken by a decade of discoveries, beginning with X-rays and radioactivity, which were still not fully understood. Moreover, Max Planck of Berlin, who now sat across from Marie at the table, had argued that energy did not flow in a continuous stream, as had long been assumed, but was chopped up into discrete packets that he and Einstein called "quanta." This new view of energy remained highly controversial, which was the reason it topped the Solvay Council agenda. The week promised to spark intense discussions, all moderated by Hendrik Antoon Lorentz, a University of Leiden professor whose investigations into the nature and behavior of light had earned him the 1902 Nobel Prize in Physics.

First Solvay Council, Brussels, 1911: Mme. Curie confers at table with Henri Poincaré; Jean Perrin leans his head on his hand beside her. Ernest Rutherford stands directly behind her, with James Jeans to his right and, to his left, Heike Kamerlingh Onnes, Albert Einstein, and Paul Langevin. Founder Ernest Solvay is seated third from left, with chairman Hendrik Lorentz to his left. Max Planck is standing second from left.

Lorentz, at fifty-eight a senior member of the group, had earned a reputation for disarming, skillful diplomacy. "As things stand," he acknowledged at the start of the meeting, physicists "currently have the feeling that we have reached an impasse. The old theories have proved themselves less and less capable of illuminating the dark shadows that seem to surround us."

Into this darkness, Lorentz said, the "beautiful hypothesis" first proposed by Planck and later applied by Einstein "has come as a precious ray of light" to reveal "quite unexpected horizons." Even those who viewed the quantum idea with "a measure of distrust"—and Lorentz counted himself one of them—"must nonetheless recognize its importance and its potential."

The intellectual conference in Brussels ended on November 3. On the fourth, a damning article appeared in the Paris daily *le Journal*, headlined "A Story of Love: Madame Curie and Professor Langevin." Other papers seized on the gossip and embellished it. When the front page of *le Petit Journal* trumpeted "A Romance in a Laboratory: The Love Affair of Mme. Curie and M. Langevin," an outraged Jacques Curie leapt to Marie's defense. "In the name of the Curie family," he wrote in a letter to the editor, "it may be useful to say that my sister-in-law has always been in her private life as perfect and remarkable as she has been distinguished from the scientific and general point of view." She had brought only happiness to her husband—"impossible to imagine two natures more perfectly matched"—and to Dr. Curie, who passed the last years of his life in her home. Jacques himself felt as profoundly attached to her as to a true sister. "I believe I can say that she and I have full confidence in each other, coming from the bottoms of our hearts, and that nothing in the future will ever part us."

Marie had only just refuted the most scurrilous of the press accounts—and won a printed apology from a guilty reporter—when she received word on November 7, her forty-fourth birthday, that she was to be awarded a second Nobel Prize, this time in chemistry, for the discovery of radium and polonium, the determination of radium's atomic weight, and its purification in the metallic state.

Navigating a path between scandal and tribute, she wrote privately to Svante Arrhenius, the 1903 laureate in chemistry who now occupied a commanding position on the Nobel Committee. She thanked him for the great honor. However, she said, she feared the rumors circulating about her might disturb the solemnity of the December ceremony, and she wanted his opinion as to whether she should attend it or stay away. Arrhenius assured her that no one in Stockholm believed the lies in the French press. He urged her to plan her trip without delay.

As these letters traveled to and from Sweden, the set of letters stolen from the rue Banquier apartment at last appeared in print in the weekly *l'Oeuvre*. On November 23, the day of the publication, a mob gathered outside Marie's house in Sceaux, denouncing the "foreign woman," the "husband stealer." André Debierne came to her rescue. He shepherded Marie and the children to a safe haven at the Paris apartment of their friends Émile and Marguerite Borel.

Paul Langevin felt compelled to challenge the editor of *l'Oeuvre*, Gustave Téry, to a duel. Téry's article had branded Paul "a cad and a scoundrel"—insults that required a specific response from a gentleman of the *Belle Époque*. Paul conscripted mathematician Paul Painlevé to be his second, then went to purchase a pistol. Langevin and Téry met on the field of honor in the Bois de Vincennes at 11:00 a.m. on November 26, but neither man fired a shot. As Téry later stated in the pages of *l'Oeuvre*, "I had scruples about depriving French science of a precious brain."

The published letters and the much-publicized duel caused Svante Arrhenius to reconsider Mme. Curie's visit to Stockholm. "All my colleagues have told me it is preferable that you not come here on December 10," he wrote her on December 1. "I therefore beg you to stay in France."

"The action which you advise," Marie replied, "would appear to be a grave error on my part. In fact the prize has been awarded for the discovery of Radium and Polonium. I believe that there is no connection between my scientific work and the facts of private life." She could not see why her acceptance of this supreme recognition for her research "should be influenced by libel and slander concerning my private life." She believed many people shared her view of the situation, and she felt "very saddened" that Arrhenius himself did not.

The strong tone of her letter masked the fact that she was drained by the ongoing ordeal, and also ill with a vaguely diagnosed kidney ailment. As before, when she shared the physics prize in 1903, she did not feel well enough to travel, but she would not back down. Both Bronya and Irène accompanied her to Stockholm. Although she had come to receive her second Nobel Prize, she now got her first taste of official Nobel festivities. She attended the formal ceremony at the Royal Academy of Music in a simple dress that stood out against a sea of tuxedos and white ties, bowing as King Gustaf handed her the eighteen-karat gold medal and ornate diploma in their leather cases. At the banquet that followed in the Hall of Mirrors of the Grand Hôtel, she drank champagne and dined on lobster with pickled winter vegetables and Jerusalem artichoke purée, guinea hen with porcini mushrooms and lingonberries, poached pearl onions with parsley roots and velouté sauce, topped off by mandarin and white chocolate mousse on a cinnamon-spiced cake with raspberry marmalade and fresh raspberries. In *December*.

"Radioactivity is a very young science," she reminded the king and other diners in her brief banquet remarks. "It is a child whom I saw being born, and whom I have helped, with all my strength, to raise. The child has grown, it has become beautiful."

Her acceptance lecture, delivered the next day at the Nobel Institute, perforce recalled Pierre and their shared years in the shed. As Pierre always did, she named the other scientists who had made essential contributions to the study of radioactivity: Henri Becquerel, Ernest Rutherford, Frederick Soddy, William Ramsay, André Debierne. She credited Rutherford in particular with establishing "a backbone for the new science, in the form of a very precise theory [i.e., transmutation and the pattern of radioactive decay] admirably suited to the study of the phenomena." She also clarified her role in the unfolding of events around radium: "The history of the discovery and isolation of this substance furnished proof of my hypothesis, which holds that radioactivity is an atomic property of matter and can provide a method for finding new elements." She proudly described every step that had led to radium's secure placement on Mendeleev's periodic table.

Unfortunately, she said, she had not achieved the same for polonium despite her mighty efforts over the years. She blamed the difficulty on the minuscule proportion of polonium in pitchblende. Radium was rare enough, but polonium far rarer—by a factor of five thousand.

Rarity severely limited the radioactivity researcher, Mme. Curie pointed out. Most labs involved in this work had only a milligram of radium salts at their disposal, at most a gram—a quantity valued at 400,000 francs (about double the monetary portion of her Nobel Prize). "And yet we have methods of measuring so perfect and so sensitive that we are able to know very exactly the small quantities of radium we are using . . . to within 1 percent of a thousandth of a milligram." In other words, "we have here an entirely separate kind of chemistry . . . which we might well call the chemistry of the imponderable."

Chapter Thirteen

HERTHA (Carbon)

———

MARIE RETURNED FROM Stockholm exhausted, overextended, and still afflicted by the kidney infection. Between her physical pain and her emotional pain, she could no longer tolerate the daily commute between Sceaux and the rue Cuvier. Nor could she face the jeers and gossip that continued to unsettle her once-peaceful home. She had found a large apartment within walking distance of the lab, at 36 quai de Béthune, on the Île Saint-Louis, and planned to move her family there in early January 1912. Although she hated to leave the area she associated so closely with Pierre, she had reason to believe she would soon rejoin him there as a permanent resident of the Sceaux cemetery. These morbid thoughts held a firm basis in Marie's reality: at forty-four, she was already two years past the age at which her mother had died.

Shortly before the planned move to central Paris, however, on December 29, 1911, a feverish Marie collapsed and was taken by ambulance to a hospital. She spent the next four weeks as an inpatient, then convalesced further at home. For the first time since being named a Sorbonne professor, she was unable to teach her course. Day-to-day direction of the Curie lab fell, of necessity, to André Debierne.

It was the worst possible moment to accommodate the new chemist, Margarethe von Wrangell, but Marie had promised her a placement. Mlle. von Wrangell, a daughter of Russian nobility, had lived a cultured life of leisure in Estonia before deciding, near age thirty, to study science in Germany. Within six years at Tübingen she completed her advanced degree in chemistry, then returned to Estonia to work at an agricultural experiment station near Tartu. She arrived in Paris fresh from a formative radioactivity immersion with William Ramsay in London. But instead of receiving Mme. Curie's personal attention, as she had expected, she continued her thorium research under the supervision of M. Debierne.

Later that winter, Marie was again hospitalized. This time she underwent surgery for the removal of kidney lesions. She emerged even weaker than before. She could not predict when she would again feel well enough to resume her teaching, her cryogenic collaboration with Heike Kamerlingh Onnes, or any of her other work. Fortunately, she had finished purifying and weighing the required quantity of radium chloride for the all-important radium standard the previous August. Her final product weighed 21.99 milligrams, or slightly more than the twenty she had promised. She sealed the material in a tiny, thin-walled glass tube, just 32 millimeters long (an inch and a quarter). The standard was now in safe keeping at the lab, where she hoped it would remain. After all, she had provided the starting material and performed the labor herself. Even if she were reimbursed for the expense, she would still wish to keep the standard near her. She had said as much to Ernest Rutherford when she saw him in Brussels at the Solvay Council in the fall. As the two of them talked privately, late into the night, she told him the simple truth: not only did she wish to monitor the standard's activity over time, which was reason enough for holding on to it, but she also felt sentimentally attached to it. Rutherford argued convincingly that the standard could not be held in private hands. Their fellow members of the International Commission for the Radium Standard would never allow it, he said. But no formal decision had yet been reached.

"Mme. Curie is rather a difficult person to deal with," Rutherford wrote Bertram Boltwood soon after that tête-à-tête. "She has the advantages and at the same time the disadvantages of being a woman."

It seemed to Marie that Paris was as central a repository as could be desired for the international standard. The city was already home to the international kilogram and the international metre standards.

Why should the international radium standard break with that tradition? Moreover, radioactivity researchers would have access to so-called "secondary standards," based on the international standard and distributed around the world. She was willing to prepare the secondaries as well, just as soon as she regained her strength.

Meanwhile she bore her months of prescribed inactivity as obediently as possible. She passed hours watching the silkworms that Irène and Ève raised in glass jars on the windowsills. In a letter to her niece Hanna, who had been sent back to Warsaw on account of Aunt Manya's illness, she said that the silkworms' incessant industry sometimes made her fancy herself one of them. Admittedly, Marie wrote, she was not at all suited for silkworms' work, but, like them, she always strove with determination toward a goal. Like them, she said, she had never had the least assurance of knowing her goal was true, or whether she could reach it, regardless of the effort expended. "I did those things because something obliged me to, just as the silkworm is compelled to spin its cocoon," she told Hanna. "The poor caterpillar has no choice but to take up the task, and if she fails to complete it, she dies without metamorphosing, without recompense of any kind." These musings led her to conclude, "Each of us must spin our own cocoon, dear Hania, without asking why, or to what end."

BY NOW STEFAN MEYER, who served as secretary of the international commission, had produced a standard for convenient use at the Radium Institute in Vienna. Meyer's access to the rich resources of the St. Joachimsthal uranium mine enabled him to make an in-house laboratory standard comparable in size to Mme. Curie's international radium standard. It seemed reasonable, therefore, to arrange an actual comparison of the two. "I go over on Saturday next to Paris to take part in the comparison of the Vienna and Paris standards," Rutherford informed Boltwood on March 18, 1912. "Mme. Curie is, I understand, not well enough to take part, but Debierne will be her representative. She was anxious to postpone the meeting, but Meyer felt that it would delay matters very much, and there was some doubt as to how long the delay would be before she could hope to attend. I have not much doubt but that the two standards will be found in very good agreement, but it will be a devil of a mess if they are not."

Marie had known Stefan Meyer since the early days of radioactivity, when she and Pierre sent him samples of their new elements for study. Over the years he had helped her acquire several additional tons of pitchblende from St. Joachimsthal, including the material for her recently attempted spectroscopy of polonium. The last time she had seen him, at the 1910 meeting of the standards committee, she noticed that his fingertips, like hers, bore the scars of a long, close association with radium. In Meyer's case, the damage wrought more consequential effects: he regretted he could no longer play the bass viol.

Rutherford and Meyer stopped at Mme. Curie's apartment on the day of the comparison for an informal luncheon with her family and a few other members of the international commission. Frederick Soddy, now head of a radioactivity research group in Glasgow, came, as did Otto Hahn of Berlin and Egon von Schweidler, Meyer's colleague at Vienna. After the meal and a short meeting, Debierne led the men to Gabriel Lippmann's laboratory at the Sorbonne, where the two standards were compared according to their emission of gamma rays. The agreement between them proved, in Rutherford's words, "as close as the measurements were possible within a limited time, at any rate, to 1 point in 300, and may have been closer."

In April, seeking the curative effects of fresh air and anonymity, Marie used her sister's name, "Mme. Dluska," to rent a small house in the countryside. Her girls rode the train there on weekends or when school breaks allowed. In May she learned that Barbara Ayrton, daughter of her friend, physicist Hertha Ayrton, had been arrested and held at London's Holloway Gaol for militant activism in the suffragist cause. Barbara had since been freed, but three leaders of her organization, the Women's Social and Political Union, remained prisoners. Hertha, herself a veteran protest-marcher, implored Marie to please sign an international petition in support of their release. "I am a member of the association whose leaders are now in prison," Hertha said in her letter, "and I know those leaders personally and look on them as persons of the utmost nobility of mind and greatness of purpose. May I hope that you, whose name is a household word among us and will command respect from the whole civilized world, will help to procure this act of justice for those who are devoting their lives to procuring the enfranchisement of women in England?"

Marie had not acted politically since leaving Poland and was at present concealing her identity from scandal-mongers and other privacy-invaders. Nevertheless she lent her name to Hertha's cause, saying, "I have great confidence in your judgment and am convinced that your sympathies are well placed." Moved by the suffragists' struggle, she said she wished for their success. The petition did succeed in its immediate purpose, though voting rights still eluded the petitioners.

Bronya, Józef, and Helena had each visited Paris to support Marie through one or another of her recent difficulties. May brought other callers from Warsaw to her door: A contingent of Polish scientists tried to entice her back to her homeland. They offered her the directorship of a new radium institute, to be built for her in her native city. "Deign, most honored madame," their official letter requested, "to transport your splendid scientific activity to our country and our capital."

They had arrived at least two years too late. Marie could not relinquish control of the radium institute being built here in her daughters' native city. Nor did she want to. But neither could she reject this entreaty from her countrymen, especially as it honored radium. So she accepted the new responsibility, promising to guide the institute from afar. As for the on-site direction, she delegated two assistants from the Curie lab who had Polish roots and were equal to the task, Jan Kazimierz Danysz and Ludwig Wertenstein.

In June, still unwell and pitifully thin, Marie took the healing waters at Thonon-les-Bains, in the French Alps near Lake Geneva, chaperoned by her friend Alice Chavannes. At the spa she went by the name "Mme. Sklodowska." "I am pursuing my water cure in a tranquil place," she wrote to Ellen Gleditsch, who had recently completed her degree and returned to Norway. "My health is improving very slowly." Ellen sent her replies, as instructed, to the Curie lab, care of André Debierne, who forwarded Madame's mail and guarded the secret of her whereabouts. "Unfortunately," Marie told Ellen in a subsequent letter, "I am always in pain, and cannot write to you at length."

Now Hertha Ayrton, who had been begging Marie for well over a year to spend a month or two with her in England, finally prevailed. Hertha boasted impeccable bona fides as a nurse: She had tended her mother, her sister, and her husband through protracted illnesses, and regularly cared for suffragists leaving prison half starved by hunger strikes. Twelve years older than Marie, she fully intended to restore her younger friend's health by summer's end.

Hertha chose not to host Marie at her London home in Norfolk Square, as the place was under constant police surveillance because of the militant suffragists sheltered there. Instead, Hertha met "Mme. Sklodowska" at Dover in late July and escorted her to an old Hampshire mill house she had reserved at Highcliffe-on-Sea. The place stood right on the Channel coast, with nothing but some woods between its garden and the shore. Hertha spoke fluent French, thanks to her lifelong closeness with cousins in Paris, and when Irène and Ève arrived, she found a local governess to tutor them in English. Irène, nearly fifteen and already well advanced in her language study, attempted both Dickens's *David Copperfield* and Coleridge's "Rime of the Ancient Mariner." She also read French poetry with her mother, copying out certain verses for discussion in depth. The water at Highcliffe-on-Sea was cold for bathing, but not much colder than the sea in Brittany.

The two-month idyll afforded Hertha and Marie a rare leisure to converse about their work. Hertha maintained a physics laboratory in her home, and relied on just one assistant, a Mr. Greenslade, for her research and demonstrations. Of her several patented inventions, she was best known for eliminating the noisy hiss and flicker of the carbon arc lamps that lined London's avenues. Hertha had successfully penetrated the Institution of Electrical Engineers, where she remained the only woman among more than three thousand men. The Royal Society, on the other hand, had refused to make her

Hertha Ayrton in her home laboratory

a fellow on the grounds that she was a wife. Even so, the society did award her its Hughes medal in 1906 and allowed her to present her own papers at meetings—two noteworthy female firsts.

While Marie had merely modified her given name to sound more French when she moved to France, Hertha had taken her name from the title of a poem, "Hertha" by Algernon Charles Swinburne. She believed Hertha suited her better than Phoebe, the name her parents gave her, or Sarah, the middle name by which her family had addressed her since childhood. Her husband had called her by the tender nickname B.G., for Beautiful Genius.

The names Marie had been called in taunts spawned by the Langevin affair branded her a foreigner, a home-wrecker, and even a Jew. She was in fact Catholic, both by birth and by upbringing, but had lost her faith after her mother's death. Hertha, who never hid the truth of her Jewish background, had likewise abandoned religious observance at an early age. Both women had struggled and sacrificed to gain a university education, and continued, each in her own way, to help other women advance. If Mrs. Ayrton felt any remorse about her illegal involvement in the fight for voting rights, it was only because she questioned the effect of her efforts. "I often think very sadly," she wrote to one of her intimates, "that perhaps I should have been more useful to the Cause if I had devoted myself to my own special work as Madame Curie has done."

Their time together passed pleasantly, but it did not bring Marie relief from pain. She returned to Paris still unable to take control of her laboratory. Margarethe von Wrangell, frustrated by the lack of contact with Mme. Curie, went back to Estonia and found employment at another agricultural experiment station, this time near Tallinn.

Irén Götz also left the Curie lab, partly on account of Madame's health problems, and partly because of her own. "The illness that kept me from finishing my work in your laboratory also prevented me from thanking you in person," she wrote from Budapest in late September. "Feeling better now, I hasten to acknowledge everything you did for me by admitting me to your laboratory, which allowed me to complete my studies and to learn things that I could not have learned anywhere else." She had profited handsomely during her eighteen months in residence, even coauthored a report with Jan Kazimierz Danysz in *Le Radium* about beta radiation. Still, she rued the fact that the "regrettable circumstances of last winter" had deprived her of Madame's unique counsel.

Viewed in a positive light, the frequent departures of female students and workers from the Curie lab translated into an increase in the number of trained women scientists employed elsewhere. Marie's first student assistant, Eugénie Feytis, was now in Zurich, on sabbatical from Sèvres to pursue research in magnetism with Pierre Weiss at the Swiss Federal Institute of Technology. Ellen Gleditsch was teaching the first university course in radioactivity ever offered anywhere in Norway. She had also written another of her popular magazine articles, this one a profile of Mme. Curie. Sybil Leslie, at Manchester, had impressed Rutherford enough to merit mention in one of his letters to Boltwood: "Miss Leslie is comparing accurately the diffusion constants of the thorium and actinium emanations." Eva Ramstedt had been hired at the Nobel Institute for physics and chemistry, to work under Svante Arrhenius. Harriet Brooks had two children now, two-year-old Barbara Anne and a son, Charles Roger Pitcher, born the previous January.

By the autumn of 1912, all of the young women Marie had accepted into the Curie lab had moved on.

Chapter Fourteen

SUZANNE
(Platinum and Iridium)

———

SUZANNE VEIL, a native Parisian of twenty-six years, arrived at the Sorbonne Annex at the start of the fall term in 1912, expecting to gain experience in radioactivity under Madame's guidance. Faced with the reality of the director's ongoing absence, however, Mlle. Veil sidled instead into the chemistry laboratory of Georges Urbain, an expert on the rare earths—a group of fifteen closely related metallic elements.

For the first time since the establishment of the *Laboratoire Curie*, no women were to be seen working there, not even Marie. Her convalescence stretched through the autumn and into the winter.

When at last she set out, on the morning of February 21, 1913, on her first official errand after the long illness, she traveled across town with André Debierne to deposit her precious radium standard at the *Bureau International des Poids et Mesures* (BIPM). The bureau, which was home to the Metre and the Kilogram, stood next door to the school at Sèvres where she had taught her first physics classes. The sight of the towering chestnut trees in the Parc de Saint-Cloud welcomed her back to familiar ground. At the same time, the choice of the bureau as repository appeased the foreign members of the radium-standard committee, who had bristled at the thought of allowing the international standard to be kept in Mme. Curie's lab. To them, the BIPM represented neutral territory with global recognition. And although it was certainly true that the bureau wielded worldwide authority, Marie had finessed the radium arrangement through her personal rapport with its deputy director, Charles-Édouard Guillaume, an old and trusted friend of Pierre's.

Guillaume stipulated that the radium standard be stored *unofficially*, because the bureau owned no in-house expertise in radioactivity. Guillaume himself specialized in metal alloys, such as the combination of platinum and iridium that composed the international prototypes of the Metre and Kilogram. These two priceless objects, along with six exact copies of each, resided in an underground vault with a triple-locked, double-iron door. The radium standard would be kept in a separate safe, located one floor up from the platinum-iridium prototypes, among other duplicates of them. Of course, the BIPM could not be held responsible in the event that any misfortune, such as fire or theft, befell the radium standard. Therefore, the less said about the matter, the better. As agreed upon in advance, no publicity or ceremony attended the moment when the radium standard, sealed in its small glass tube and packed in a protective metal box, left Marie's hands to take its newly assigned place on a shelf inside the designated safe, to which Guillaume held the only key.

Marie's compensation for the expense of fabricating the standard had likewise come about unofficially, through personal contacts. When the question of cost first arose, at the 1910 congress in Belgium, Ernest Rutherford had thought of appealing to government stakeholders for funding. Going that route, however, would have invited the intrusion of government bureaucrats into the committee's doings, and so Rutherford urged the members to seek other sources of financial support. In 1912 Frederick Soddy's wealthy father-in-law, the industrial chemist George Beilby of Glasgow, agreed to reimburse Mme. Curie for the radium invested in her standard. In return, she publicly thanked "Dr. and Mme. Beilby" in an article for the *Journal de physique*.

Responsibility for producing the secondary standards fell to the Radium Institute of Vienna, which could tap the rich pitchblende source of the St. Joachimsthal mine. The secondaries were to contain smaller quantities of radium than the international standard, and thus would be less costly to produce, but comparably pure. Once they passed muster, the secondaries would be distributed for local use in Germany, the United Kingdom, the United States, and beyond— wherever radioactivity research took root.

Having duly delivered the radium standard to the BIPM in February, Marie returned to retrieve it just a few months later, in May. The first of the secondary standards had arrived from Vienna, bound for the National Physical Laboratory in England. She performed the necessary tests at her lab, which she had lately taken to calling the Paris Laboratory of Radioactivity. The comparison tests involved quantifying the secondary's radioactive emissions over a period of weeks by measuring its gamma radiation via electroscope. She then signed a certificate of authenticity and took the international standard back to Sèvres, where it would remain until she needed to borrow it again.

The Curie lab had been acting as a de facto measurement service almost since the discovery of radium. Pierre and Marie had personally set the standards for the French radium industry when they gave Armet de Lisle's *Sels de Radium* the Curie seal of approval. In recent years Marie had also been called upon to certify the output of two additional French radium factories, as well as the radioactive products employed at various hospitals and doctors' offices. Providing this service struck her as such a crucial activity that she had written it into her plans for the new radium institute now under construction. Meanwhile she carried on the measurement work as professionally and efficiently as possible. Each time a request reached the Curie lab, she complied in order to validate the radioactive sample in question—and never to advance one individual's or institution's interests over another's. The fee she charged in exchange for the time she or André Debierne devoted to these efforts defrayed some of their other expenses. Under no circumstances did she permit anyone whose products earned a Curie-lab certificate to tout that fact in commercial advertisements.

Despite her scrupulous handling of these matters, she fell afoul of Sorbonne rector Louis Liard, who accused her of commercialism and of acting without proper authorization from the Faculty of Science. Marie retorted that her certification was sought because of her personal expertise, not her affiliation with the university, and furthermore that she did the work for the public good, accepting payment only for the time spent. As for the granting of signed certificates, which particularly rankled Liard, she argued that she could not guarantee the specifics of her detailed analysis with a simple handshake. A formal investigation headed by Georges Urbain exonerated her of any misdeed.

Slowly she revived her research, beginning with a review of the extreme-low-temperature studies she had undertaken with Heike Kamerlingh Onnes. Although their collaboration at Leyden had ended in July 1911, they had held off publication all this time in the hope of extending their experiments. "But," they explained in their 1913 coauthored report, "as the continuation of the work has been prevented so far by the long indisposition of one of us, we thought it best not to wait any longer in publishing our results." As far as they could tell, extreme cold exerted no effect on radium's characteristic behavior.

———

MADAME'S PROTÉGÉE Ellen Gleditsch, while striving to awaken Norwegian national interest in radium, suffered a devastating series of personal losses in the early months of 1913. First her mother, Petra Brigitte Hansen Gleditsch, died of tuberculosis in February. Then her brother August returned from engineering school in Germany, sick and dying of the same disease. Within weeks, their father, Karl Kristian Gleditsch, also succumbed to tuberculosis.

At the time of the family's great hardship, all of Ellen's surviving siblings had reached maturity except for the youngest one, eleven-year-old Kristian. Ellen, now thirty-two, instantly assumed responsibility for the boy's care. Her sister Birgit, the second oldest, might have been willing to assist her but had married a missionary and gone with him to China. Ellen found a new apartment in the capital, large enough for herself, Kristian, and their brother Adler, age twenty. On her meager salary as an instructor, she accepted the burdens of

Hiking in Switzerland: Miss Manley (the governess)
in front with Ève Curie and Hans Einstein; at back,
Albert Einstein, Marie Curie, and Irène Curie.

marriage and motherhood without the benefit of a husband. "Fortunately," Marie consoled her in a sympathy note, "I know you are strong and courageous and able to overcome your suffering by thinking of those who need you."

Marie tested her own strength and courage that summer by taking Irène and Ève on a walking tour through the Alpine valleys of the Engadin in southeastern Switzerland. There they hiked daily with Albert Einstein and his son Hans. She had not seen Einstein since meeting him at the Solvay Council in Brussels, where they had had time to become friendly just before she became a victim of the press. "I am so enraged by the base manner in which the public is presently daring to concern itself with you," he wrote her from Prague in November 1911, "that I absolutely must give vent to this feeling." Dismissing the lurid news reports as an attempt by the "rabble" to "satiate its lust for sensationalism," he said, "I am impelled to tell you how much I have come to admire your intellect, your drive, and your honesty, and that I consider myself lucky to have made your personal acquaintance in Brussels." He was "certainly happy," he added, "that we have such personages among us as you, and Langevin too, real people with whom one feels privileged to be in contact."

On their mountain walks the two physicists lagged behind the youngsters, parsing topics of mutual interest in a mixture of German and French. Ève later recalled howling with laughter at odd snippets of the grown-up conversation. She swore she and Irène heard Einstein tell their mother, "You understand, what I need to know is exactly what happens to the passengers in an elevator when it falls into emptiness."

Although the recent past had been a fallow period for Marie, she had followed the research activities of all her colleagues. Einstein was expanding his theory of relativity. Rutherford's experiments with alpha particles had revealed the structure of the atom. To his and others' great surprise, the positively charged particles clumped together at the center, ringed at a distance by the negatively charged electrons. Rutherford called the central mass the "nucleus," a term he borrowed from cell biology.

The atom's exterior electrons conducted the ordinary business of chemistry, namely combining with other atoms. Radioactivity happened inside the nucleus, where, at any moment, the sudden, violent expulsion of a nuclear particle would convert an atom of one element into an atom of another.

These infinitely small-scale events exerted vast influence. According to the latest calculations, the heat released in radioactive decay warmed the whole Earth from within. Geologists had thought the Earth to be steadily cooling since the time of its formation, and feared its frozen demise within a few millennia. But radioactivity projected the planet's life expectancy far into the future, and by the same token pushed its antiquity farther back into the past. Rocks dated by the new technique of measuring the proportion of uranium to lead showed themselves to be amazingly ancient. The estimated age of the Earth jumped from a few tens of millions of years to several billion.

By 1913 the radioactivity community recognized three distinct series, or "families," of radioelements, each with its own pedigree. Radium belonged to the uranium ancestral line, or "uranium family," distinct from the "thorium family" and the "actinium family." A tally of all the decay products in all three lines of descent added up to thirty-four radioelements, all vying for barely one-third that number of vacant places on the periodic table between uranium, where radioactivity began, and lead, where radioactive decay ended. The superabundance of intervening substances threatened to topple the table. In an attempt to restore order, Frederick Soddy and others pointed out that several radioelements resembled one another too closely to be considered discrete elements. They were more like variations on a theme. Although they differed in their radioactivity, with different emissions and different half-lives, they behaved identically in chemical reactions. Once mixed together, they could not be parted from one another by any known means. Even their spectra looked identical. Since they were chemically the same, Soddy argued, they could share the same location on the periodic table. He was holding forth about these "chemically non-separable radioelements" one evening at the home of his in-laws, the Beilbys, when another dinner guest, the Glaswegian medical doctor and novelist Margaret Todd, coined a word to describe such chemical look-alikes. "Isotope," she suggested, from the Greek for "same place."

Isotopes relieved the immediate space problem on the periodic table. They also hinted at a deeper truth underlying the table's organization. Mendeleev had relied on atomic weight as his guiding principle, arranging the elements in ascending order accordingly. But the fact that each isotope bore a slightly different atomic weight suggested that something other than weight—something even more fundamental—must serve as an element's key defining feature. No one could say what that distinction might be.

To account for the weight disparities between isotopes, scientists envisioned a form of neutral ballast residing in the nucleus. A little more or less of this as yet unknown stuff would spell the difference between thorium, say, at atomic weight 232, and its isotope radio-thorium, at 228.

In the new light of isotopes, Marie could see why her 1902 and 1907 figures for radium's atomic weight did not quite match. The tiny discrepancy indicated no lack of technique on her part, no lapse in skill, but merely a different assortment of isotopes each time. Since radium now shared its place on the periodic table with several chemical twins, such as mesothorium 1 and thorium X, the atomic weight of radium in any given trial must represent an average or a median between the heaviest and lightest isotopes in the mix.*

Marie had played no active part in the recent discoveries about atomic structure and the nature of isotopes. Yet, in a sense, she had enabled them all. She was the one who had pursued Becquerel's uranic rays, found them in new places, given them a new name, and recognized radioactivity as an atomic property of certain elements. Suspecting the presence of polonium and radium in pitchblende, she had pried them from rock dust by the force of her will. Everything she had done to secure radium's place on the periodic table strengthened her own position as standard-bearer for her discovery.

---

*The International Union of Pure and Applied Chemistry currently recognizes thirty-three isotopes of radium.

Chapter Fifteen

MAURICE (Ionium)

———

PARTLY OUT OF RESPECT for Mme. Curie, partly as a show
of friendship, and partly "to smooth things over" in case her feelings had been bruised in discussions of the radium standard, Ernest
Rutherford entreated the University of Manchester to grant her an
honorary degree. He had intended to make the presentation at the
formal opening of his new laboratory extension, in February 1912, but
she was too ill then to travel, and the diploma was simply mailed to
her at home. The offer of another honorary doctorate, this one tendered by the University of Birmingham and timed to coincide with
a meeting of the British Association for the Advancement of Science,
took Marie to England in the late summer of 1913. From the London home of her friend Hertha Ayrton, she sent birthday greetings
with "best kisses and deepest love" to her elder daughter, *"ma chère
grande Irène,"* who turned sixteen on September 12. "I embrace you
with all my heart, my child," she wrote, "and I enclose a procedure
for constructing an ellipse that you may not have met with before."

Irène and Ève were vacationing with their uncle Jacques and his
family in Langogne. On the morning of her birthday, Irène spruced
up her cousin Maurice's bicycle, then set off with him and Ève on a
forty-five-kilometer ride. Eight-year-old Ève, who felt very much
the little-sister tagalong, described the day-long excursion and picnic
in a note to her mother, signed *"Microscopique bébé."*

Marie deemed the investiture ceremony in Birmingham a strange
rite. "They dressed me in a beautiful red robe with green trim," she

told Irène, "just like my companions in misery—that is to say the other scholars receiving the doctor's degree." She said she used the solemn occasion "to take note of the laws and customs of the University." She had spoken English with everyone, including Messieurs Rutherford and Soddy, "whom you met at our house."

While Rutherford was entertaining Mme. Curie, word reached him of her protégée Ellen Gleditsch. "I have a piece of news that will interest you," wrote Bertram Boltwood from the Hague, where he was winding up his European summer travels. "Mlle. Gleditsch has written that she has a fellowship of the American Scandinavian Foundation (I never heard of it before!) and wishes to come and work with me in New Haven!! What do you think of that? I have written to her and tried to ward her off, but as the letter was necessarily

Seated between Sir Oliver Lodge (left) and Dr. Gilbert Barling, Mme. Curie was honored at Birmingham with Robert W. Wood (standing, left), H. A. Lorentz, and Svante Arrhenius.

delayed in forwarding to me, I am afraid she will be in New York before I get there. Tell Mrs. Rutherford that a silver fruit dish will make a very nice wedding present!!!"

Boltwood made no jokes about matrimony in his response to Ellen herself:

> In regard to your carrying on scientific work in radioactivity at my laboratory in New Haven, I may state that I should be very glad to have you do so provided that the facilities I can offer are sufficient to make you feel that it is worth your while . . . I am not at all well supplied with radioactive preparations such as you must have become accustomed to have at hand during your experience in Madame Curie's laboratory in Paris.

Undeterred, Ellen sailed for America. Over the summer she had prepared her two younger brothers to manage the little household without her, because, as she made them see, she desperately needed another stint in an established radioactivity laboratory abroad. Her home institution, the Royal Frederick University, provided neither fellow faculty members with similar expertise nor sufficient equipment for her experiments. Even with the backing of the American-Scandinavian Foundation, however, she needed to take the initiative to secure a placement with a well-known radioactivist in the United States. "No woman," Theodore Lyman at Harvard wrote in reply to one of Ellen's inquiries, "has so far set foot in the physics laboratory of Harvard." The same held true at Yale, she was certain, but at least Boltwood did not bar the door.

A representative of the American–Scandinavian Foundation greeted its first woman fellow at dockside in New York. A reporter for the *New York Press* hailed her arrival with predictable disbelief: "That pretty little woman with the large brown eyes and the soft and smiling lips [is] one of the most remarkable feminine scientific investigators? Impossible! Where are her glasses, her severity, her aspect of independence? What have such words as radioactivity and gamma rays to do with such sweet lips?"

———

WHEN MARIE and her daughters reunited in Paris that fall, they made room for cousin Maurice Curie in their apartment on the quai de Béthune. Now twenty-four, and a graduate student in chemistry at the Sorbonne, Maurice proved an ideal older brother surrogate for Irène and Ève. Marie also found a place for him in her laboratory as a special attaché, and trained him to shoulder some of the certification work.

In addition to her own resumed duties at the Curie lab, Marie renewed her regular inspections of the new radium institute rising on the rue des Nourrices. Her stepped-up pace of activity cost her a relapse of kidney trouble, but she rallied quickly enough to attend the second Solvay Council, held in Brussels in October. While conferees at the first such meeting, in 1911, had addressed theories of radiation and quanta, the 1913 agenda concerned the structure of matter. Radioactivity and quantum theory had opened the atom to scrutiny. The select gathering attracted nearly all of the original Solvay participants, along with several new ones, including J. J. Thomson of the Cavendish Laboratory and German physicist Max von Laue, who had recently shown X-rays to be a very short-wavelength variety of light. Once again Marie stood out in the group photo as the only woman.

After Brussels she traveled to Warsaw for the dedication of the new radium institute there, headed by Jan Kazimierz Danysz, her former assistant. Danysz had hesitated to leave Paris, he confessed at the time of his departure, and above all her laboratory, where he had spent four happy years. Here in Warsaw, he consoled himself, he was still working for her.

The return to her native city allowed Marie to stroll the banks of the Vistula and visit the cemetery where her parents and sister Zofia lay buried. A heavy air of oppression still cast its pall. "This poor country," she reflected in a letter home, "massacred by an absurd and barbarous domination, really does a great deal to defend its moral and intellectual life." At a banquet in her honor, she embraced the aged but still lively Jadwiga Sikorska, headmistress of the girls' school she and Helena had attended. At the Museum of Industry and Agriculture, where Marie had acquired her taste for experimentation, she lectured on radioactivity to a large audience—in Polish—for the first time.

Her own long-promised Radium Institute of Paris still looked like a muddy construction site throughout the wet winter of 1913–1914. At her lab in the Annex, aided by her nephew and André Debierne, she continued to address the needs of researchers, doctors, geologists, prospectors—of anyone and everyone who required reliable assessments of the radioactivity of mineral deposits, thermal waters, and medical materials.

One day Monsieur Petit, who had been a lab assistant at the industrial school, came to tell Marie that the old hangar behind the building on the rue Lhomond was about to be torn down to make room for a new wing. She immediately went with him to see the abandoned shed one last time. After a decade's absence, she found Pierre's blackboard still standing in its remembered place, still bearing a few faint lines in his hand that no one had bothered to erase.

———

AT YALE, Bertram Boltwood challenged Ellen Gleditsch to determine the half-life of radium. He had previously estimated the figure to be two thousand years, but Rutherford, using a different technique, claimed a considerably shorter period of sixteen hundred years. Could Miss Gleditsch settle the four-century discrepancy?

Boltwood was an early convert to the "new alchemy" of radioactive transmutation. The natural occurrence of uranium with radium strongly suggested that the one decayed into the other, and he tested the idea by measuring the ratio of radium to uranium in minerals from different parts of the world. He found these to be wonderfully consistent—about three-tenths of a microgram of radium to every gram of uranium. Seeking further evidence of consanguinity, he watched over a piece of pure uranium in his lab, hoping to witness a radium birth, but thirteen months passed between 1904 and 1905 with no sign of such an event.

Boltwood's failure to "grow" radium from uranium led him to suspect a missing link in the family line. Some other radioelement must arise from uranium breakdown and give rise in turn to radium. In 1907 he discovered "the direct parent of radium." He named it ionium, in recognition of its potent ability to ionize (that is, excite) the air around it to conduct electric charge—a hallmark of radioactivity.

Ionium generated radium, which then transformed into emanation. Boltwood reached his value for the radium half-life by clocking the speed at which the radioactive gas accumulated. Rutherford, on the other hand, had counted the alpha particles that a sample of radium emitted per second by watching them scintillate in green on a zinc sulfide screen, and then extrapolated.

Ellen judged Boltwood's approach the superior method and thought she could improve its accuracy. As she had done in the Curie lab while studying the radium-uranium ratio, she repeated Boltwood's experiments, but with more materials than he had tested and under stricter constraints. She began with a sample of uraninite from North Carolina. She ground 110 grams of it into a fine powder, then dissolved the powder in warm nitric acid. Some two dozen steps later in her recipe, she was left with a residue of pure ionium, which she dissolved in dilute hydrochloric acid and sealed in a glass bulb. After eight days she boiled off the emanation and measured its volume. The quantity of emanation in the bulb indicated the amount of radium now present in the solution. At the end of six weeks, she again boiled off and remeasured. Now she had an inkling of the rate at which radium decayed. She would test several more times at different intervals in order to tabulate and interpret her results. Meanwhile she started a parallel procedure with a piece of the radioactive mineral cleveite from Norway. Already she sensed that Boltwood's half-life estimate of two thousand years was too high.

As she acclimated to her new surroundings, she compared this second experience of research abroad to her first. Not only did she find herself the only woman in the Sloane Laboratory, but she rarely saw another woman anywhere on the Yale campus. Moreover, Boltwood, a bachelor, never hosted her at his home on Sundays as Mme. Curie had done in Paris.

Ellen sought social connections by venturing to some of the nearby women's colleges, such as Vassar in New York and Smith in Massachusetts. The students there enjoyed her guest lectures on radioactivity. Aside from her good command of English and complete mastery of her subject, she knew how to frame her remarks and modulate her voice to keep an audience engaged. One invitation led to another. In February the president of the American Chemical Society, Theodore W. Richards, asked her to visit Harvard. By this point Richards's colleague in the Harvard physics department, Theodore Lyman, had softened his attitude toward Ellen Gleditsch, and expressed himself willing to let her be the first woman to set foot in his lab. But she was embedded now with Boltwood at Yale.

The experiments were going well, with only occasional mishaps. She prepared a third ionium solution, this time from Norwegian bröggerite. She tested her original North Carolina solution again at eleven weeks and then at fourteen, and the cleveite according to a different schedule.

In April Ellen took the train to Washington, DC, to hear Ernest Rutherford inaugurate a prestigious lecture series at the National Academy of Sciences. He had recently been knighted in the 1914 New Year's honors. Unfortunately, Ellen observed, Sir Ernest lost a number of interested listeners at the academy through his regrettable habit of swallowing the ends of sentences. She seized a moment with him after his talk to ask about his thoughts on her current research. After all, she might take issue with one of his published results, and she felt obliged to apprise him of that possibility. Without reservation, he encouraged her to forge ahead.

Ultimately, by employing Boltwood's method, Ellen arrived at Rutherford's figure of sixteen hundred years for the half-life of radium.* This made everyone happy. In June, Smith College awarded

---

*Sixteen hundred years is the value still accepted today as the half-life of radium-226.

her an honorary doctoral degree for "exceptional intellectual attainments . . . in this new and important science."

On her way home in June, Ellen stopped in England to spend a few days at Manchester with Curie-lab alumna Jadwiga Szmidt. Together they learned the shocking news that student revolutionaries in Sarajevo had assassinated Archduke Franz Ferdinand of Austria and his wife, Duchess Sophie of Hohenberg.

The diplomatic fallout from the June 28, 1914, event reverberated across the continent. Although anxiety at the threat of war quickly pervaded private life, people tried to carry on as usual. In mid-July, for example, Irène and Ève set off, as planned, with Joséphine the cook and Walcia the Polish governess, to the family's favorite seaside haunt at l'Arcouest on the Brittany coast. Marie promised she would follow the moment she finished transferring certain contents of the Curie lab in the Annex to the new *Institut du Radium*, which was finally ready to open its doors.

"Dear Irène, dear Ève," she wrote on Saturday, August 1, still stuck in the city. "Things seem to be getting worse: we expect the mobilization from one minute to the next. I don't know when I will be able to leave, and communications may be cut off. Don't panic. Be calm and courageous. If war does not break out, I'll come and join you on Monday. If it does, then I'll stay here and send for you as soon as possible. You and I, Irène, we shall try to make ourselves useful."

Part Three

The Radium Institute: Curie Laboratory

1 Rue Pierre-Curie

Odette's appearance, to one who knew her real age and anticipated seeing a very old woman, seemed to oppose the laws of chronology even more miraculously than the persistence of radium defies the laws of nature. If I failed to recognize her at first, it was not because she had changed but because she had not.

—Marcel Proust, *In Search of Lost Time*

No, no army follows at your heels,
No saintly voices call your name,
But your heart's earnest, burning zeal
Rivals every pyre's flame.

—"Ode to Madame Curie," by Maurice Rostand,
recited by Sarah Bernhardt at the Paris Opéra,
April 28, 1921

Chapter Sixteen

IRÈNE (Lead)

———

WHEN GERMANY, under the complex system of alliances then in effect, declared war on France on August 3, 1914, Hertha Ayrton invited Marie and her daughters to shelter with her in London. Marie, however, chose to stay in Paris, determined to contribute, somehow, to the war effort. While she strategized, her girls remained in relative safety at the seashore. She counseled the excitable Irène to be patient, to build her physical strength for whatever might be required of her, and to "stand in my place by your little sister."

Irène chafed to leave l'Arcouest. On the brink of her seventeenth birthday, she, too, wanted to defend France. She missed her mother. She wondered whether she would be able to begin studies at the Sorbonne as they had planned. Her mood rose and fell in response to news reports and rumors. One day she felt overcome by torpor, as though trapped in a bad dream, but then the next day found her calm, resolute, as she hoped she would continue. On the morning of August 3, she went to Paimpol with the Perrins, to see the father of her friends Francis and Aline off to army service. Every family at the rail station looked heavyhearted, all bidding what might be a final goodbye to someone beloved. Then a military band began to play, and Irène watched the music work its effect on the crowd. People raised their heads, their hats, their hands. Handkerchiefs, which only moments earlier had been dabbing at tears, waved in the air in time to the march as the train departed.

The mobilization emptied the Curie lab, as every able-bodied man became a soldier. André Debierne was in Normandy, awaiting orders, and Maurice Curie at Vincennes. Even Marie's former assistant, Jan Kazimierz Danysz, rushed back from the Warsaw Radium Institute to join the French infantry. With communications slowed or cut off, Marie heard nothing from her family in Poland.

She imagined she could make herself most useful in tending the wounded. For the first time in the history of warfare, it would be possible to see inside an injured body with X-rays, to locate the lead bullets and fragments of bombs that lodged there. X-rays, which had entered medical diagnosis almost immediately upon their discovery by physicist Wilhelm Roentgen in 1895, were now employed at most major hospitals and also some private clinics. But many more X-ray machines, along with the skilled professionals to operate them, would be needed to meet the looming demand.

Marie conferred with Dr. Claudius Regaud, director of the medical half of the *Institut du Radium*, and Dr. Antoine Béclère, a practitioner and authority on the diagnostic and therapeutic uses of radiation. Dr. Béclère had instituted the country's first instructional course on X-rays for medical doctors in 1897, and was still offering it, every other year for three weeks. Just as Marie had coined the term "radio-activity," he had introduced "radiology" to describe the exterior examination of the body's interior. Now wearing the uniform of an army major despite his nearly sixty years, Dr. Béclère kept his hands, which bore the effects of a decade's daily exposure to X-rays, concealed in a pair of gray gloves.

The flood of wounded men arriving from the front quickly filled the huge military hospital, *Val de Grâce*, and now men were being taken to other facilities lacking the latest X-ray equipment. Mobile X-ray units that could travel between these points—and even to the zones where troops were deployed—seemed the obvious solution. In fact, the army was already outfitting a few trucks for this purpose. Perceiving the enormity of the need, Marie moved to procure and equip such vehicles herself with the aid of other private citizens.

Her first radiology car, a boxy Renault, came as a gift from the women of the French Red Cross—*l'Union des Femmes de France.* The vehicle accommodated all the needed apparatus, yet was small and agile enough to navigate the narrowest streets in the city. Unfortunately, Marie did not yet know how to drive, but she would learn. Till then she could sit in the back seat with the elaborate and delicate blown-glass bulb that beamed the X-rays. She would need a chauffeur anyway, to help her unload, set up, and then reload the crates of equipment, and perhaps to assist her, or, in places with no electricity, to keep the engine running as a power source. As she gathered materials and reacquainted herself with human anatomy, she conscripted other cars for transformation into similar *voitures radiologiques,* including the personal limousines of the Marquise de Ganay and the Princess Murat.

By the end of August the German forces had advanced to within thirty miles of Paris. As troops defended the city, the government relocated to Bordeaux, about three hundred miles southwest of the capital. Concerned for the safety of her cache of radium, Marie used her connections to secure it a temporary haven. The president of the republic, Raymond Poincaré, was the brother of her recently deceased Sorbonne colleague Henri Poincaré. Soon an official decree proclaimed that "the radium in the possession of Mme. Curie, professor of the faculty of sciences of Paris, constitutes a national asset of great value," and must be held at Bordeaux "for the duration of the war."

Marie transported it herself. The gram of radium, worth about a million francs, existed as various preparations of bromides and chlorides filling numerous glass tubes. These she packed with lead protection in an overnight bag. Case and cargo together, she guessed, weighed twenty kilos (about forty-five pounds). She carried no other luggage when she left the lab on the afternoon of September 3, wearing her black cloak, and boarded a train reserved for government personnel. En route, she saw that the roads visible from the tracks were packed with carloads of fleeing Parisians. At Bordeaux her traveling companion, a ministry employee, directed her to a room for the night in a private apartment. In the morning she deposited the radium at a bank vault, then rode a troop train back to Paris. She was the only civilian passenger. As hours elapsed with many unexplained stops and lengthy delays, a soldier shared his bread with her.

At home she found a letter from Irène. Anxious l'Arcouest locals had accused the family's devoted cook and governess, with whom the girls conversed in Polish, of speaking German. Of *being* German. Irène herself was suspect. "I am going to give you some examples: they say that I am a German spy. They say also that, when I leave in the morning with a little pail to gather some blackberries, I am carrying things to eat for a hidden German spy. They say also that I am a German man disguised as a woman, etc." This gossip reached her in a roundabout way, yet found its mark. How could anyone mistake her for a foreigner, she sputtered, "when I am so profoundly French and I love France more than anything? I can't keep from crying every time I think of it, so I will stop in order to keep this letter legible."

"Don't take these things too much to heart," Marie replied on September 6, "but do your best to clarify the situation for those around you." What a pity, she said, that her recent travels had taken her so far from Paris yet no nearer to her daughters. She reminded Irène of the caring adults, including the local mayor, whom she could call on for help at any time. "I pray that our reunion will come soon," Marie added. "It pains me to think that I probably won't be able to embrace you on your birthday next week. Rest assured that my thoughts will not leave you that day, and that I'm determined to recoup all the time I've been deprived of you . . . *Au revoir, ma chérie, Ta mère*."

As she wrote, fighting broke out in the valley of the Marne. Marie received good intelligence on French progress from friends and relatives at the front. Throughout the month of September, some half a dozen skirmishes felled most of the officers in Jan Danysz's regiment. He and his comrades took shelter in holes they dug, covered with branches and dirt. "We have put straw inside and we sleep there with our clothes on," Danysz wrote to her on October 4. "Every day we improve our kit by bringing everything that could be useful from a neighboring village: a chair, a table, a mattress, blankets—with the result that our holes become more and more habitable. The nights are getting a little cold. Fortunately, there is no rain . . . We are some hundreds of metres from the German lines, with the result that we hear gunfire or take shelling all the time. I am beginning to get used to it."

Within weeks, however, Marie learned to her great sorrow that Danysz had been killed in combat.

"One hundred thirty kilometers on foot," her nephew Maurice reported in mid-October, "sack on back, rain and mud, hardly any food, across the mass graves of Épernay, Montmirail, etc. It is unbearable."

—————

WHEN MARIE AT LAST permitted her daughters to return to the apartment on the quai de Béthune, Irène learned she could take mathematics courses at the Sorbonne and also a crash course for nurses. Classes resumed at Ève's school, too, and when she was not busy reading or practicing the piano, she knit sweaters for soldiers. Together, the girls marked war maneuvers with little flags stuck into a large wall map hanging in the dining room.

On November 1, Marie deputized Irène and the Curie lab mechanic, Louis Ragot, to accompany her to the Second Army's evacuation hospital in Creil, about twenty miles north of Paris. Ragot had a heart condition that exempted him from active duty, but Marie knew him capable of repairing anything. On arrival they carried in the boxes of supplies, which contained everything from the big glass bulb and its electrical connections to photographic plates and chemicals for developing them, as well as black curtains, in case the room where they were to work had a window. Radiology required darkness. It would take at least half of the half-hour setup time for their eyes to acclimate to the dim light. When all was ready, the atmosphere recalled the rooms where Marie had long ago attended séances with Pierre. Here, in this make-do radiology suite, she herself served as the medium, and would soon summon visions to stun the skeptical surgeons and nurses waiting impatiently to operate on the wounded.

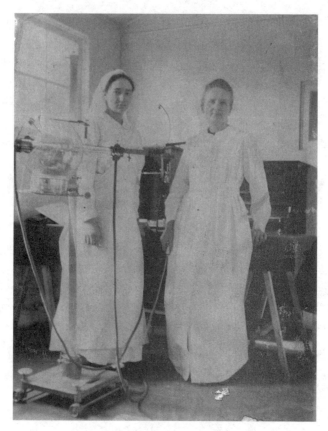

Marie and nurse Irène Curie on X-ray duty
in Hoogstade, Belgium, 1915

Most medics had seen an X-ray image at some point in their training but had gained no personal experience with the process or its
utility. Months of war had accustomed them to probing blindly for
projectiles that might have strayed far from their entry points. Such
surgery could do further damage, especially when it failed to locate
and extract the foreign object.

If the doctors looked askance at Marie's futuristic apparatus, the
casualties being carried in on stretchers feared a new assault from
it. She assured each man that being X-rayed was no more painful
than being photographed. She wheeled the glass bulb on its trolley

into place on one side of him, opposite the fluorescent screen that the rays would strike after passing through his chest, his abdomen, his thigh—wherever his injuries indicated. Then a hiss and a spark and suddenly that body part was rendered transparent on the screen. Among the familiar bones, now splintered or broken, the leaden intruder stood out starkly.

———

DR. CLAUDIUS REGAUD, or rather Lieutenant-Colonel Regaud, took charge of reorganizing the military health service, the *Service de santé*, and expanding the radiology facilities. It was imperative to operate on the wounded immediately, since transferring them to hospitals allowed time for infection to set in. Marie tried to convince the army brass that she, too, in her trusty Renault, should go wherever wounds were inflicted. But in spite of her experience and demonstrated ability, she was often thwarted by regulations regarding the presence of women near the battlefields.

"The day I leave is not fixed yet, but it can't be far off," she wrote Paul Langevin on January 1, 1915. Now that Paul had reunited with his wife, Jeanne, Marie communicated with him openly. "I have had a letter saying that the radiological car working in the Saint-Pol region has been damaged," she continued. "This means that the whole north is without any radiological service! I am taking the necessary steps to hasten my departure. I am resolved to put all my strength at the service of my adopted country, since I cannot do anything for my unfortunate native country just now, bathed as it is in blood after more than a century of suffering."

She had never recovered her full strength after her long convalescence, and still lost days at a time to recurring attacks of kidney pain. On the road, she discovered her *voiture radiologique* could cruise at twenty-five miles an hour. "Dear children," she scrawled in haste on January 20, "here we are at Amiens, where we slept. We have only burst two tires. Regards to all. Mé." Later that day she sent more news: "Arrived at Abbeville. Jean Perrin, with his car, ran into a tree. Luckily no great harm done. Continuing to Boulogne. Mé."

At first the physicists drafted into the army were sent, without much thought to their special expertise, to guard bridges and roads at strategic points, but later they were reassigned to radiology. Although X-rays illuminated the internal landscape, they yielded only a two-dimensional picture, whether instantaneously on the fluorescent screen or later on a developed photographic plate. These images were conic projections, typically pitched at odd angles. It took an observer with knowledge of physics and geometry to see past the distortions and calculate the exact location of the object to be removed. Also, a physicist accustomed to tinkering with lab machinery could usually diagnose and address any problems that arose with the X-ray source or the electrical transformer.

François Canac, who had started out at the Curie lab in 1909 as an independent researcher, wrote to tell Marie he had crossed paths with Maurice and another colleague from the Annex: "You can imagine with what joy I had these two encounters and thus reconstituted the little scientific kingdom of Paris."

Maurice reported frequently. "I should like to leave the village where I am," he wrote his aunt at the end of February. "One winds up being a ruin living amid the ruins. This hole has been so thoroughly demolished that they are running a new train line through it in order to reprovision Verdun. I would give my blanket for an hour spent at the window of the quai de Béthune."

Everywhere Marie ventured—to Verdun, to Reims, to Calais—she won converts to X-ray examination. Sometimes an entire operation would be performed "under the rays," enabling the surgeon to follow the course of his forceps by looking at the radioscopic screen. More often, Marie would trace the image that appeared on the screen, using paper and markers she brought with her, and the sketch would serve as the surgeon's guide. The best images were the radiographs captured on photographic plates, but these had first to be developed—in tubs and with solutions that formed a standard part of the five hundred pounds of portable radiology equipment. The radiograph provided not only the sharpest possible image but also a permanent record of the original wound. Subsequent radiographs over the ensuing days or weeks could track the progress of recovery from a fracture or other injury. It was not unusual for Marie to leave a site with the promise to return and install a quasi-permanent radiology post. Then she needed to make good on the promise, with support from the privately funded agency dedicated to helping disabled soldiers, the *Patronage des blessés*, or the army's own *Service de santé*.

In addition to performing X-ray duty, Mme. Curie's *voiture radiologique* functioned as a van for moving laboratory apparatus from the rue Cuvier to the *Institut du Radium* in the newly named rue Pierre-Curie. One morning in March 1915, looking ahead to the day when research would resume in the new space, Marie visited the flower market for attractive shrubs and bulbs. As she planted, the bombardment of Paris began. She continued the task even as "a few shells fell in the vicinity."

In April, on the way home from an X-ray mission in Forges-les-Eaux, her driver swerved and the car overturned as it fell into a ditch.

Mme. Curie in her *voiture radiologique*, 1917

Temporarily buried under some of the heavy packing cases, Marie emerged bruised and bloodied but not, by her account, seriously hurt.

Irène, who often accompanied her mother, managed to keep up her Sorbonne studies through the first year of the war. She earned her *certificat* in mathematics, with distinction, around the same time she received her nursing diploma. She was to start physics in the fall, but meanwhile her scholastic achievements enabled her to spend the summer as a full-time radiology assistant.

Unlike her mother, who simply added a Red Cross armband to her everyday clothes, Irène wore an ankle-length, all-white nurse's uniform and tied back her hair under a veil. All through July and August, while her cousin Maurice lamented "the monotony of the trench . . . the lines of barbed wire, the huts, the mines, the shells, the bullets," she met their victims in the hospital wards of France and Belgium. Her eighteenth birthday, September 12, 1915, found her in Hoogstade, Belgium, bunking with other nurses and supervising X-ray examinations.

"I enjoyed my birthday," she wrote her mother the following day, "save for the fact that you weren't with me, *ma douce chérie*. First I found the apron that I had accused you of mischievously pinching. Then I took a radiograph of a hand with four pieces of shrapnel large enough for me to localize and which will be extracted today." Later she attended an afternoon soccer match and spent the evening listening to a little concert. "After all that, I went to sleep in the tent under a beautiful starry sky."

Chapter Seventeen

MARTHE (Chlorine)

WHILE MME. CURIE traveled through France in her *voiture radiologique* during the first year of the war, the enemy introduced a new way of felling combatants: by fouling the air they breathed. An onslaught of lethal chlorine gas released at Ypres in the early spring of 1915 took Allied soldiers unawares and choked many of them to death. In May, barely a fortnight after this event, Madame's English friend Hertha Ayrton conceived the idea for an ingenious hand-operated fan that could both repel and clear poisonous vapors from the trenches. She made the prototype "Ayrton Fan" in her private laboratory in London, based on her previous observations of vortices in water. The device, weighing only a few ounces, consisted of paper flaps, or blades, connected by leather hinges and mounted at the end of a wooden stick. It bore a certain resemblance to an oversized flyswatter. Mrs. Ayrton admitted that she "laughed aloud at the simplicity of the solution." But others laughed, too. Indeed, although she offered her invention free of charge to the War Office, she was ignored and even ridiculed for months before being invited to demonstrate the fan at Chatham in September 1915. An official report enthusiastically endorsed the flapper's surprising utility, but did not, unfortunately, spur the armed forces to begin mass production.

"Tomorrow morning," Maurice Curie wrote to his aunt in mid-November, "they will close us in a room and the atmosphere will be charged with asphyxiating gases, tear gases, etc. We are trying on the new protective masks."

In January 1916, Mrs. Ayrton dispatched her assistant, C. E. Greenslade, to the front to demonstrate the fully realized fan, in which the paper blades had been replaced by waterproof canvas stiffened with cane. Greenslade showed how incoming clouds of gas could be driven back by beating the fan against the parapet of the trench. Then he voluntarily entered a dugout in which a cannister of deadly gas had been exploded. Wearing no mask or other protective gear, he walked through the pit, pushing the gas before him by flapping the fan along the dirt floor. He emerged minutes later into a crowd of men stupefied to see him still alive.

For the purpose of clearing gas from trenches, observers judged the Ayrton fan twice as effective as the "sprayers" in use by the Allied armies. Even so, four more months passed before the first shipment of five thousand fans arrived in France.

"I suppose," Mrs. Ayrton fumed to her friend and biographer, Evelyn Sharp, "if I had invented something to destroy life instead of saving it, it would get taken up at once as a military proposition!"

By now Mme. Curie's war experience had convinced her that X-ray examination had to be offered to the sick as well as to the wounded, and extended from the obvious injuries to every part of the body. At one hospital she visited, she met a soldier who had a fractured pelvis. Already bedridden for weeks, he was not expected to live. X-ray examination, though slow and very painful on account of his poor condition, revealed a piece of shrapnel above his knee, inside a large pocket of pus. On the very day that surgeons removed this foreign

The Ayrton Fan

body and drained the infected area, the youth's overall health started to rebound, and soon his broken bones knit back together.

In order to recruit more of the desperately needed X-ray operators to man the mobile units and staff the radiology outposts, Mme. Curie approached professors, engineers, and university students who were either free from military service or stationed conveniently where she needed them. Too often frustrated by losing these men to military transfer orders, she decided to train women as well, and tapped her network of former assistants. Suzanne Veil, who had exited the Curie lab almost as soon as she entered it in 1913 due to the director's protracted absence, now got to work under her at last. Because Mlle. Veil already understood the physics of X-ray generation and the calculus required for localization, she learned quickly. Soon she qualified to train others.

Eugénie Feytis, who studied physics in Madame's class and later taught the subject at Sèvres, recommended several of the school's recent alumnae as likely radiologists. Eugénie herself, however, was unable to participate in the X-ray work. In the summer of 1913, at the end of her studies in Switzerland with Pierre Weiss, she had married Aimé Cotton, a physicist twelve years her senior, whom she knew through the Sunday gatherings of scientists in the Curies' garden. As she had seen Marie do, Eugénie kept working after she wed, both as a teacher at Sèvres and in pursuit of her doctoral degree at the Sorbonne. Her own widowed mother, Émilie Menant Feytis, came to live in the apartment below the Cottons, where she helped care for little Eugène, born on July 21, 1914, about two weeks before the war broke out. Although Aimé's age excused him from active duty, he collaborated with his friend Pierre Weiss on methods and instruments for pinpointing the location of enemy artillery by sound-ranging, then trained groups of physics students to employ their techniques. Aimé was with the army near Verdun in February 1916, coaching and encouraging these specialists in the trenches, when Eugénie gave birth prematurely to their second son, Paul. Had an incubator been available, then perhaps, despite the wartime privations and the winter cold, the infant might have lived.

———

THE RADIUM STASHED for safekeeping at Bordeaux at the start of the war returned to Paris in 1916 under orders from medical authorities in the military. Dr. Regaud and Dr. Béclère felt confident that radioactivity, already proven a boon in the treatment of cancer, could also ameliorate the vicious scars, adhesions, arthritis, and neuritis that followed battle wounds. Like a medicinal plant with balm in its leaves, roots, and seeds, radium released a variety of potential remedies, including penetrating gamma rays and gaseous emanation. A gram of radium, properly milked in the new Curie lab at the Radium Institute, would yield a steady supply of emanation, which could be bottled and distributed in therapeutic doses to soothe a battalion of injured soldiers.

"Having no assistants," Mme. Curie wrote of this effort, "I had, for a long time, to prepare these emanation bulbs alone, and their preparation is very delicate."

To provide care for many thousands, more radium and more personnel would be needed. Production at *Sels du Radium* had ceased during the war for lack of manpower, but when Mme. Curie sounded its owner, Émile Armet de Lisle, she found him amenable to reopening if skilled workers could be found. On June 22, 1916, she appealed to Ellen Gleditsch in neutral Norway. "I write to you today to ask you if you would like to come to France in order to work in the laboratory at the Armet de Lisle factory. Mr. Armet would be very happy to have you here and to restart the work of the factory, and I think he will offer you good conditions."

While waiting for Ellen's reply, Marie applied for and obtained her driver's license. Irène completed her Sorbonne *certificat* in physics, again with distinction, and went to work as a nurse in Montereau. In July Marie sent Ève to l'Arcouest with Henriette Perrin and her children, Aline and Francis. Ève wrote home chastising herself for losing her wristwatch on the train, a mishap that made her feel she was not yet mature enough (at age eleven) to own precious things. Soon she forgot the trinket in the grip of adventures one could have at l'Arcouest, such as running the rowboat aground on a sandbar and needing to wait hours for the sea to float it free, or setting off with three companions on an all-day hiking excursion that covered thirty or forty kilometers and included an elaborate picnic lunch. Even without her mother's on-site supervision, however, Ève could not escape math lessons, and found fifteen-year-old Francis Perrin a stern professor. "Today I solved a first-degree equation all by myself," she wrote *ma douce Mé,* "and without errors, and then we also did other exercises in algebra, and he made me memorize three formulas by heart."

———

ELLEN GLEDITSCH had recently secured the newly created post of docent in radiochemistry at her university when she received Mme. Curie's invitation. Mindful of the war dangers, to say nothing of her concern and ongoing responsibility for her youngest brother, she recognized the need and agreed to go. Her route to Paris took her across the North Sea and the English Channel—two bodies of water now menaced by German submarines.

All Allied attempts at anti-submarine warfare had failed thus far, but Ernest Rutherford in England and Paul Langevin in France were determined to develop means for detecting these vessels by sound. Paul had begun experimenting with acoustic apparatus in March 1915 in his laboratory at the *École Municipale de Physique et de Chimie Industrielles,* where he once studied under Pierre Curie. A promising first trial on the Seine in April 1916 prompted the Navy Department to transfer Langevin's group to Toulon, on the Côte d'Azur. Now he was testing an active sonar model in actual sea conditions, somewhat hampered by the unusually cold and rainy weather.

When Ellen got to Paris in July, she found parts of the city utterly changed by war and other parts exactly as she remembered them. Much of daily life went on as usual, so that those with means and leisure could still dine out in restaurants or see films at the cinema. Everyone Ellen had met at the old Curie lab was engaged in war work of some kind. May Sybil Leslie, a college chemistry instructor in Wales before the onset of hostilities, was now chemist-in-charge of explosives at a government munitions factory near Liverpool. She supervised a staff of six in her department but assigned the most dangerous jobs to herself. Eva Ramstedt, who stayed in close touch with Ellen, had left the Nobel Institute to teach radiology at the University of Stockholm. The two Scandinavians were coauthoring a book about radioactivity based on Mme. Curie's lectures and publications.

At the factory in Nogent-sur-Marne, Ellen resumed some of the first procedures she had learned as a novice radioactivist, albeit with new precautions gained from long practice. Radiation's effect on the skin, as she and Eva pointed out in their soon-to-be-published text, "is considerably weaker if the preparation is held at a distance, and especially if the rays have first to pass a metal screen. Therefore, when you are carrying radium salts with you, these should be kept in lead capsules." They advised using tongs to handle vials or bottles of radium salts in the lab, and also switching the tongs periodically from one hand to the other. In general, they counseled, one should never hold anything radioactive in the same position for any length of time.

In the fall Mme. Curie issued an open call for women to enroll in a new course of study at the Radium Institute, aimed at training female X-ray operators, or *manipulatrices radiologistes*. She believed that a reasonably intelligent young woman with only an elementary education could, if properly motivated, master the knowledge and skills for providing X-ray examination in about two months. Applicants needed to be native or naturalized French citizens, in good health, and at least twenty-one years of age. Irène had only just turned nineteen when the first session got under way in October 1916, but she was a teacher in the program, not a student. At the same time, she continued her own studies at the Sorbonne, concentrating next on chemistry.

Each class of twenty aspiring *manipulatrices* submitted to a packed six-week curriculum of electricity, measurement of electric current and potential, electromagnetic induction, theory and function of X-ray tubes and valves, and methods of examination by radioscopy and radiography. Anatomy lessons with Dr. Nicole Girard-Mangin, the sole female physician in the French armed services, took place at the newly established Edith Cavell Hospital.

In theory, Marie was preparing these women to serve as aides to physicians, though she thought several of them quite capable of working independently. Doubtless the war would accord them that opportunity.

———————

AS PROMISED, Ellen Gleditsch oversaw production at *Sels du Radium* for six months. Heading back to Norway shortly before Christmas, she retraced her route across the Channel, through England, and over the North Sea. Although the waters still roiled with submarines, Paul Langevin had hit on a new detection approach based on the piezoelectric effect discovered by Jacques and Pierre Curie. He would use the piezoelectric properties of quartz to transform sound waves generated by moving ships into electrical oscillations that could be amplified for playing into headphones. He also planned to employ the quartz as a transmitter, to send out supersonic waves and, he hoped, thereby locate the underwater enemy. In April 1917, when the United States entered the war, Langevin prepared to share his thinking with English and American naval authorities, to avoid anyone's duplication of efforts. The first such inter-Allied meeting took place in Paris in May. When Ernest Rutherford arrived, Marie toured him through the new Radium Institute. Hers was the only building in the city surrounded by walls of stacked sandbags, installed by the army to safeguard the Emanation Service from the shock of falling shells.

"Mme. Curie gave me some Lab. tea," Sir Ernest wrote home to his wife, Mary Newton Rutherford. "She is looking rather grey and worn and tired. She is very much occupied with radiology work, both direct and for training others."

Various Sèvres connections led Mme. Curie to a new assistant. Although the thirty-two-year-old Marthe Klein came with no prior laboratory experience, she had studied physics under Eugénie Feytis and finished first in her class. Since her graduation in 1908, Mlle. Klein had taught mathematics at girls' schools in several cities, from Saumur to Marseilles to Bordeaux, hoping all the while for an appointment in Paris, where she could pursue her own higher education at the Sorbonne. A scholarship took her to the University of Cambridge for the 1913–1914 academic year, but after that one taste of advanced study, which was all she could afford, she returned to France and resumed teaching at Bordeaux. In October 1916, when her fellow Sèvrienne Lucie Blanquies became director of the *lycée Racine*, Marthe filled Lucie's old slot at Versailles. Finally she was in proximity to the university. In the summer of 1917, she began volunteering at the *Institut du Radium*, lecturing to a new wave of *manipulatrices*.

Marthe immediately proved herself indispensable. Marie, dreading Marthe's September return to the classroom at Versailles, appealed to the undersecretary of state in charge of the *Service de santé* to keep her employed as an instructor of radiology for the duration of the war.

The need for *manipulatrices* grew along with the number of X-ray outposts and *voitures radiologiques* patrolling the battlefront. Marie personally outfitted eighteen such cars. The army versions of these so-called *petites Curies* grew to the size of trucks. They not only carried X-ray equipment but also functioned as rolling darkrooms for developing the radiographs. The number of soldiers examined by X-ray was nearing one million.

———

THE EARLY MONTHS of 1918 saw an intense bombardment of Paris but left the Curie home and lab intact. "One is so overcome by the precarious living conditions," Marie wrote to Ellen Gleditsch in January, "that one doesn't have a moment of satisfactory peace of mind to gather thoughts about one's friends." Nevertheless, she acceded to a request from the Italian government to visit that summer and assess the country's natural resources in radioactive materials. She left in mid-July for a month in northern Italy.

Irène spent part of July in Langogne with the Jacques Curie family. After going over some of the formulas in the text by Lord Rayleigh (John William Strutt, the 1904 Nobel Prize winner in physics), she wrote her mother, "I then reviewed, with a little help from Maurice's copy of Nernst, the calculation of the speed of diffusion of ions of both signs."

Ève, at l'Arcouest awaiting her mother, was content to sweeten her coffee with saccharine instead of sugar, to do without green vegetables, and to accept jam as the only available form of fruit. In one of her letters, she described a convoy of fifty-three sailboats, accompanied by torpedo boats, heading for England in the hope of coming back with coal.

July in Toulon found Paul Langevin's group completing successful demonstrations of their active sonar device based on the piezoelectric effect. Paul wrote a detailed technical report, which he delivered at the next Inter-Allied Conference on Submarine Detection, held in Paris in October 1918. Days later the war ended, rendering his design instantly obsolete.

"It is going to be so pleasant," Curie lab *préparateur* Fernand Holweck wrote to Marie on November 12, "not to speak constantly of submarines, grenades and torpedoes. Radioactive projectiles are much more *sympathiques*."

Marie had stuck to her work throughout the war, but when a fusillade of guns signaled the signing of the armistice, she dropped everything to run, together with Marthe Klein, in search of a French flag to fly over the Radium Institute. Other patriots had beaten them to the shops, however, and snatched up every last one. Undaunted, Marie bought bolts of red, white, and blue fabric, and enlisted Mme. Bardinet, who kept the laboratory clean, to sew some banners, which they hung from the large windows.

The next morning Marie and Marthe rode through the city-wide victory celebration in the original *voiture radiologique*. At the Place de la Concorde, crowds blocked the old Renault for several minutes. A dozen jubilant celebrants climbed aboard its roof and fenders, to be carried along as though on a parade float. People sang the Allied anthems. Church bells rang out everywhere.

Eugénie Feytis Cotton observed the festivities from the open window of her apartment. "I stood listening for a long time to the bells of Sèvres," she wrote, "whose sound rose to our wooded hills. And because a tiny new life had recently announced itself inside me, these bells of peace produced a profound emotion."

Chapter Eighteen

MADELEINE

(Radioneon)

———

"TO HATE THE very idea of war," Marie wrote after the Great War ended, "it ought to be sufficient to see once what I have seen so often through these years: men and boys brought to the advanced ambulance in a mixture of mud and blood, many of them dying of their injuries, many others recovering but slowly through months of pain and suffering."

Peace freed her to fight disease itself. The Emanation Service she had initiated in 1916 for the relief of battlefield wounds now promised "*curiethérapie*" to a wide, desirous public. Dr. Claudius Regaud, her confrere at the head of the Pasteur Pavilion of the *Institut du Radium*, shared her commitment to this cause. Relieved of his military duties, Dr. Regaud resumed both his research and active treatment of cancer with X-radiation and radioactivity. Beginning early in 1919, he could be seen riding his bicycle to hospitals, carrying ampoules of emanation to his patients.

Mme. Curie's gram of radium, lately deemed "a national asset of great value" to the French government, was a fount of vital medicine. As soon as it reverted to her personal possession, she officially deeded her radium to the *Institut*. In a report stressing the importance of radium and the need for more of it, she called for the expansion of the institute into a national center complete with an industrial-scale factory. She reiterated the same imperative in another report, coauthored with Dr. Regaud. Ideally their factory would be located outside the city, with hangars for storage of minerals and chemical works for the extraction of radioelements, using continuously modernized techniques to retrieve the largest quantities of material at the lowest possible cost. It seemed to Marie a truism that, in order to study the radioelements, "we need to manufacture them." Meanwhile she maintained her very close connection to the factory owned by Émile Armet de Lisle, where some of her staff, now honorably discharged from the armed forces, were returning to work.

The end of the war saw no end to the need for more *manipulatrices* in civilian and veterans' hospitals. Marie continued to train new X-ray technicians, including one class consisting entirely of American soldiers. In addition to herself, her daughter Irène, Suzanne Veil, and Marthe Klein, her all-female faculty now included twenty-year-old chemist Madeleine Monin. The unusually gifted Mlle. Monin had joined the *Institut* in 1917, immediately after completing the equivalent of three years of university science preparation in one. As the recipient of a Curie-Carnegie scholarship of 3,400 francs for the 1918–1919 academic year, she was mastering the dosimetry for the Emanation Service.

Radium emanation, the substance bringing solace to so many, seemed to merit a more distinctive—less derivative—name. William Ramsay liked "niton," from the Latin word for "shining," because emanation glowed in the dark, and because "niton" sounded similar to the names he had given other noble gases. Mme. Curie suggested "radioneon," with an emphasis on emanation's radioactivity. Any official judgment on nomenclature, however, awaited the reorganization of professional societies disbanded during the war.

Under the terms of the Treaty of Versailles, signed June 28, 1919, Poland regained territory long controlled by Austria and Germany. Marshal Józef Pilsudski, who headed the military force known as the Polish Legions, took power in Warsaw on the day of the armistice and declared Poland a sovereign nation. Now Pilsudski was pushing his country's border eastward as well, successfully battling Russia's weakened Red Army. "I had lived," Marie realized, "though I had scarcely expected it, to see the reparation of more than a century of injustice that had been done to Poland, my native country, and that had kept her in slavery, her territories and people divided among her enemies."

"It is true that our country has paid dear for this happiness," Marie conceded in a letter to her brother, Józef, "and that it will have to pay again. But can the clouds of the present situation be compared with the bitterness and discouragement that would have crushed us if, after the war, Poland had remained in chains and divided into pieces?"

———

LIKE THE MAP of Europe, the periodic table emerged from the war informed by a new world order. Investigators had used X-ray techniques to reveal precisely how many positively charged particles existed in the nuclei of individual atoms. A clear connection existed between this tally, which Ernest Rutherford dubbed "atomic number," and atomic weight, because the elements stayed put where Mendeleev had placed them. But atomic numbers represented a feature more fundamental than atomic weight, and progressed regularly, by increments of one, from the lightest element, hydrogen (number 1), to the heaviest, uranium (number 92). All isotopes of a given element shared the same atomic number.

The new rationale pinpointed the locations of half a dozen gaps remaining on the periodic table. These six breaks in the consecutive numerical order—six lacunae to be filled by future finds—occurred at positions 43, 61, 72, 75, 85, and 87.

Element number 91, proto-actinium, had only lately come to light, in 1917, through the work of physicist Lise Meitner and chemist Otto Hahn at the Kaiser Wilhelm Institute in Berlin. At the time of discovery, Hahn was serving as an officer in the German army.

By 1919, radioactivists on both sides of the recent conflict were eager to restore peacetime relations. Some spoke of rescheduling the International Radium Congress, originally planned for the fall of 1915 in Vienna, where Mme. Curie was to have given an opening address.

"The old staff of the Radium Institute is now scattered," Stefan Meyer reported glumly from postwar Vienna. Under the harsh terms

Otto Hahn and Lise Meitner

of the peace treaty, amid shortages of food, and facing rampant infla-
tion due to the devalued Austrian currency, Meyer feared "we will
not be able to continue scientific work, if at all we may continue our
life." He had sent his two young children to live with his in-laws in
Ischl, because in Vienna, "babies older than one year do not get any
more milk, we have no coals and much less wood to provide them
a warm room."

Marie had invested most of the money from her second Nobel Prize
in war bonds and loans. She had tried to donate her Nobel medals,
too, in response to the government's request for gold, but the Bank
of France refused to take them. The gold medals were still in her
keeping, the money gone. Moving forward, she would support herself
and her children on her Sorbonne salary of 12,000 francs per year. She
was neither richer nor poorer than other French scientists. Nor had
she ever craved any luxury beyond a summer vacation in the open
air. As long as her health held up, she could provide for her children.

Ève contracted double pneumonia early in 1919, and required a
lengthy hospital stay to recover. Irène received a medal for her wartime service. She had managed to complete her *licence ès sciences* at
the Sorbonne and was fulfilling her birthright—her destiny—at the

Institut du Radium, alongside her mother and her "Uncle André," as *préparatrice déléguée* in the Curie Pavilion.

Several of Marie's laboratory daughters also achieved postwar distinctions. The University of Leeds conferred an honorary Doctor of Science degree on its alumna May Sybil Leslie for her work at munitions factories in Liverpool and Wales. When that job ended, the university hired Sybil as a demonstrator in the chemistry department, and soon promoted her to assistant lecturer. Irén Götz, who had come to the Curie lab from Budapest in 1911, was now the first female professor to teach at a Hungarian university. Ellen Gleditsch, foremost among Marie's protégées, had won election to the Norwegian Academy of Science and Letters in 1917, following her six-month stay at the radium factory in Nogent-sur-Marne. She was only the second woman to be honored as an Academician in her homeland, after biologist Kristine Bonnevie.

Although the radioactivity text that Ellen wrote with Eva Ramstedt was published in Norwegian, she sent autographed copies to Mme. Curie and Ernest Rutherford as soon as the book came out. Madame simply thanked her. Sir Ernest quipped, "I regret to say that I do not know your language sufficiently to read it, but I know the subject so well that I almost persuade myself that I can."

Ellen kept in touch with her American hosts, Bertram Boltwood at Yale and Theodore Richards at Harvard, about her ongoing research into the ratios of radioelements in rocks, or, as she called it, "radiogeology." In 1919, when Richards could at last travel to Stockholm to claim his 1914 Nobel Prize in Chemistry, he mentioned Ellen's name in his public lecture, citing her as one of the scientists who had aided his comparison of ordinary lead with the various isotopes of lead produced by radioactive transmutation.

At the Royal Frederick University, Ellen worked in a small, dark, humid room in the basement of the Chemistry Laboratory. The building had been poorly constructed to begin with, in the early 1870s, and showed cracks in its foundation. Everything about the place was outmoded, from gas and water supplies to ventilation. Worst of all, and despite the name "Chemistry Laboratory," it lacked adequate lab space. Ellen, who had been lecturing on radioactivity since 1912, was able to offer a related lab course only once, during the winter of 1918, when she had so few students that she could fit them all into the small room. That same year, the university agreed to build the much-needed new laboratory, and appointed Ellen to the planning committee. In 1919 she was granted a leave of absence to gather information and ideas from research centers at foreign universities. Naturally she chose to begin her site visits in Paris, at the *Institut du Radium*.

"I shall be very happy to receive you," Marie declared on August 1. She now had ample space to welcome as many as thirty or forty researchers at a time without crowding. Her own two rooms, being centrally located, allowed students to seek her help more easily than was possible in the old Annex. Already she was amassing books and journals for an in-house library of resources related to radioactivity. Little by little, as "the tempest of the last years" abated, she would establish a laboratory "such as I wish to the memory of Pierre Curie and to the highest interest of humanity." She was pleased that her quick-study assistant Madeleine Monin, who had married engineer Henri Molinier within months of meeting him at the *Institut*, saw no reason to stop working after her wedding. Nor did Madeleine's husband object to her continuing her career.

Marie spent early August pursuing the familiar summer delights of l'Arcouest with Irène and Ève, and the end of the month with Marthe Klein at Cavalaire on the Mediterranean coast. She found the southern countryside superb, the weather beautiful, and the water warm, she told her daughters, but unfortunately her room at the *Hôtel de la Plage* was noisy. She preferred the rented castle where Marthe lived communally with her friends, all professors, in what Marie described as "picturesque disorder." She slept outdoors on the balcony attached to Marthe's room until Marthe returned to Paris, then found a quieter hotel in Cavalaire and began writing a book about the uses of radiology in wartime, *La radiologie et la guerre*.

"I often think of the year ahead of us," she wrote home to her girls, "and hope it will yield good things. I also think about each of you, of all the sweetness you bring into my life, all the pleasures and all the worries, too. You are, in truth, my great treasure, and my wish is for several more good years of our togetherness."

A week later it became clear to Marie that, once again, she would likely miss Irène's birthday. Some X-ray equipment to be installed at Marseilles by the *Patronage des blessés* was delayed in transit, and the matter might not be settled before mid-month. *"Ma chérie,"* she apologized on September 7, "if I am not with you on your birthday, know that you have my deepest love. Think about the gift you would most like to receive to mark your twenty-two years. I could never find one to equal what you give me every day, by virtue of your beautiful youth, your joy in living and working, and your affection for your mother. Dearest grown daughter, blessings from the bottom of my heart for all that you are and will become, and may you find every happiness that I could hope for you. *Ma douce chérie*, I embrace you, eager to resume our tender closeness soon."

Irène, not one to gush, wrote back to say she had bought an album to hold her collection of souvenir picture postcards and was content to consider this item her birthday gift.

Chapter Nineteen

LÉONIE (Oxygen)

MME. CURIE CREATED the novel position of laboratory secretary expressly for Léonie Razet. The young Parisian widow had lost her husband during the recent war, though not to enemy fire. Jean Pierre Razet, a radioactivist who had collaborated over the years with the Curies, with Émile Armet de Lisle, and also with Jacques Danne, was unable to serve in the army because of a pulmonary ailment. He passed the early part of the war helping to dismantle the old Curie lab and set up the new one in the *Institut du Radium*, then died of his illness in December 1916, at age thirty-two, leaving Léonie the sole support of their daughters—Suzanne, eight, and Yvonne, five.

Léonie learned quickly to distinguish between matters that merited Madame's personal attention and duties she could discharge by herself. She took over the flow of international correspondence that had begun in the first blush of the radium craze, when an American admirer had tried to give a racehorse Mme. Curie's name. Nowadays the mail brought frequent invitations to lecture at foreign universities about radioactivity, to teach X-ray courses abroad for nurses and technicians, and to visit the sites of promising mine deposits or mineral waters. Some writers pleaded for medical advice, while others simply wanted the famous scientist's autograph. Quite a few sought employment. To those applicants who lacked qualifications or trusted referrals Léonie sent neatly typed regrets. In midsummer of 1919, however, when a letter from Ellen Gleditsch suggested a

PERIODIC TABLE OF ELEMENTS

Chemical Group Block

The periodic table of the elements exposes the differences between—and similarities among—the 94 naturally occurring components of the material world, as well as the 24 (at present) artificially produced elements.

Mme. Curie (seated, center) taught a physics class at Paris's best school for aspiring female teachers, the *École Normale Supérieure d'enseignement secondaire de jeunes filles* at Sèvres. Her star pupil, Eugénie Feytis (standing right), later became the school's director. The others in this 1901 photo are mathematician Anna Cartan (seated left), Marthe Baillaud (seated right), and Madeleine Routaboul.

Marie Curie received this citation along with 200,000 francs and a gold medal naming her the 1911 Nobel laureate in chemistry.

Ernest Rutherford,
1908 Nobelist in chemistry

André Debierne,
discoverer of actinium

Bertram Boltwood, radiochemist at Yale

Henri Becquerel, 1903 Nobelist
in physics (with the Curies)

Émile Armet de Lisle,
founder of *Sels de Radium*

Dr. and Mme. Eugène Curie with their
two sons, Pierre (at right) and Jacques.

Frederick Soddy,
1921 Nobelist in chemistry

Dmitri Mendeleev,
creator of the periodic table

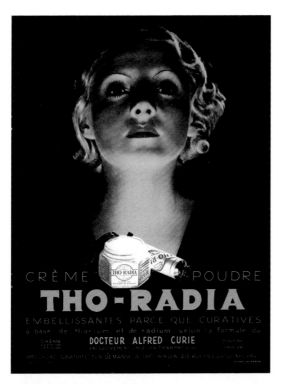

The marketers of Tho-Radia cosmetics promised youthful beauty from radioelements— a spurious claim backed by the fictitious authority of "Dr. Alfred Curie."

This 1917 class of "manipulatrices," or X-ray operators, took instruction at the Radium Institute from Marthe Klein, Irène Curie, and Suzanne Veil (seated third, fourth, and fifth from left).

The nine-going-on-ten-year-old Irène, on vacation with relatives at the shore, assured her mother that she was trying her best to master the addition of fractions.

Marie, Irène, and Ève Curie (right to left at bottom) gathered on the rocks at l'Arcouest with their friends André Debierne (at top), "Le Capitaine" Charles Seignobos (in the white hat), and other summertime visitors.

Marie stayed close to her siblings all her life and saw them as often as time allowed, as on this occasion with Kazimir and Bronya Dluski (left) and her brother, Józef Sklodowski.

Mme. Curie often stood alone—in her determination to study science, in her early grief as a widow, in the distinctions she won as a woman in a man's world. She often stood alone in the literal sense as well, staying late at her lab to continue an experiment long after colleagues and students had gone home.

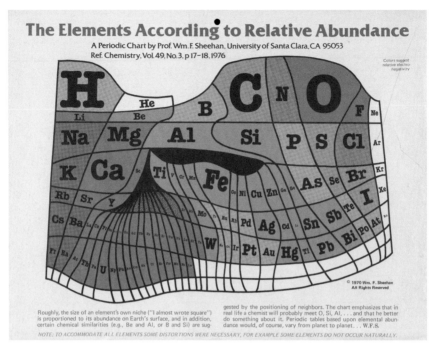

Physical chemist William F. Sheehan (1926–2008) reconfigured the periodic table to emphasize the elements' relative abundances on Earth.

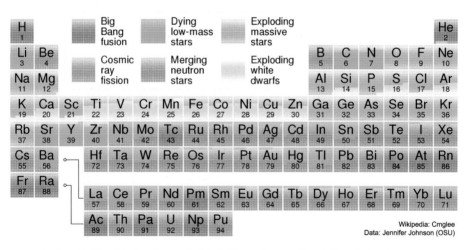

This periodic table is color-coded to indicate the cosmic origins of the elements.

candidate for the coming academic year, Mme. Curie handwrote her own positive reply.

Léonie collected a few biographical details about fellow members of the laboratory staff, and recorded them, along with brief descriptions of each one's responsibilities. These personnel files—a slim data set at Léonie's start date in 1918—grew in September 1919 to include twenty-five-year-old Renée Galabert, from Chartres, and twenty-three-year-old Sonia Slobodkine, a native of Warsaw who had just completed chemistry studies at the Sorbonne. Renée assisted Madeleine Monin in the measurement of radium and thorium products sent to the Curie lab for certification. Sonia took up the pre-purification chemical treatment of new radioactive sources.

Ellen Gleditsch returned in October with her protégée Randi Holwech, a graduate of the Norwegian Institute of Technology. Ellen had described her to Madame as "intelligent, gifted, and personable," with a demonstrated proficiency in measuring radioactivity. Randi also spoke fluent French, allowing her to engage immediately with her new associates, whereas Ellen had struggled to learn the language during her initiation into the Curie lab.

Aside from the few familiar faces of long acquaintance, Ellen saw little at the grand new institute to evoke the old lab in the Annex. Those small, overcrowded rooms, inconveniently separated by a courtyard, had paved the way to a well-designed building of stately proportions. Remarkably, the structure of the Curie Pavilion had sustained no damage from bombs or bullets, and yet the effects of the war all but echoed through the spacious halls. It would take time for the country to recover from its losses, let alone for scientific work to regain its prewar momentum.

Madame, worn down by travail and now fifty-two years old, conceded that the constant handling of radium occasionally caused her "discomfort." She was careful in the new lab to shield radioactive sources with lead—a precaution that had never crossed her mind in the days of the shed. Special screens now protected lab workers from stray radiation, and modern draft hoods carried off noxious gases. She could banish periods of "discomfort" by spending a few days away, though the dimming of her eyesight by cataracts and the humming in her ears persisted. While at work, she often stepped outdoors on the rear balcony for a breath of fresh air, and urged those around her to do the same. The balcony overlooked the rose garden. One might stand out there for a few minutes, surveying the pleasant scenery, or descend to ground level, where tables and chairs set among the flower beds invited scientists from the *Laboratoire Curie* to socialize with physicians from the *Pavillon Pasteur*.

Dr. Claudius Regaud, who had been visiting cancer patients at several clinics of the *Assistance Publique*, now delivered *curiethéra-pie* at two newly designated cancer wards, each with about twenty beds, situated in two Paris hospitals. He and Mme. Curie believed an additional dispensary belonged on the grounds of—or adjacent to—the *Institut du Radium*. In 1920 they established a foundation, the *Fondation Curie*, to accept donations and advance their plans for furthering research while treating disease with radiation.

Radium's rarity had assumed added significance with the recent increase in medical demand. The radium in Mme. Curie's lab supplied not only the physicochemical investigations under her direction but also the basic biological research on Dr. Regaud's side of the insti-tute. And it generated the emanation prized in cancer treatment. The lab's all-important Emanation Service regularly gathered precise quantities of radium-emitted gas to be encapsulated in tiny glass or platinum needles of various shapes and sizes for therapeutic insertion into any part of the human body. "Technical considerations make the employment of emanation preferable to the direct use of radium," Madame explained. Each vial came with a numerical chart that told the physician "how much of this emanation has disappeared each day, despite the fact that it is cloistered in its little glass prison."

IN MAY 1920, when an American reporter for women's magazines named Marie Mattingly Meloney got around Léonie Razet to gain an audience with Mme. Curie, the scarcity of radium dominated their conversation. "I waited a few minutes in a bare little office which might have been furnished from Grand Rapids, Michigan," Mrs. Meloney later recalled. "Then the door opened and I saw a pale, timid little woman in a black cotton dress, with the saddest face I had ever looked upon."

Marie was just back from a round of lectures and teaching in Madrid, where she had met King Alfonso XIII and his mother, Queen Maria Christina. She received Mrs. Meloney cordially.

"Her well-formed hands were rough," the reporter observed. "I noticed a characteristic, nervous little habit of rubbing the tips of her fingers over the pad of her thumb in quick succession. I learned later that working with radium had made them numb."

Having finally succeeded at meeting the long-sought object of her admiration, Mrs. Meloney struggled to match the real Mme. Curie with the preconceptions she had brought to the interview: "I had been prepared to meet a woman of the world, enriched by her own efforts and established in one of the white palaces of the Champs-Élysées or some other beautiful boulevard of Paris." Instead, "I found a simple woman, working in an inadequate laboratory and living in a simple apartment on the meager pay of a French professor." This discovery unhinged Mrs. Meloney. "Suddenly I felt like an intruder. I was struck dumb. My timidity exceeded her own. I had been a trained interrogator for twenty years, but I could not ask a single question of this gentle woman in a black cotton dress." Mme. Curie filled the awkward pause by naming the American cities where rich troves of radium resided—four grams in Baltimore, six in Denver, seven in New York, and so on. The Curie lab, in contrast, had one. Its market value, Mrs. Meloney learned, was $100,000.

"I also learned that Madame Curie's laboratory, although practically a new building, was without sufficient equipment; that the radium held there was used at that time only for extracting emanations for hospital use in cancer treatment." This seems an exaggeration, or at least a misunderstanding. Nevertheless, after two subsequent encounters, one at the institute and one at the Curie apartment on the quai de Béthune, Mrs. Meloney vowed to procure a second gram of radium with the aid of the women of America, "and in this way to enable Madame Curie to go on with her work."

———

CONTINUING HER TOUR of inspection as envoy of the Royal Frederick University's laboratory planning committee, Ellen Gleditsch left Paris for England in the spring of 1920. Her friends Eva Ramstedt and Sybil Leslie met her in Oxford. "It's the first time we are together since 1911, when we all worked in your laboratory," they told Mme. Curie via postcard. "We think of you, talk about you, and with our good memories we send our admiration and greetings."

The trio then visited Frederick Soddy in his new lab at Oxford University. Soddy had spent his undergraduate years at Oxford, before heading to McGill, where he and Ernest Rutherford uncovered "the new alchemy" of radioactive transmutation that made their reputations. His current mission at Oxford was to build a research school in radioactivity, as he had done successfully in Glasgow and Aberdeen. But he apologized for having so little to show his guests. His work had been interrupted by the war, he said, also by moving, and it would be a while before he could get things up and running.

Rutherford, too, had recently returned to his roots, leaving Manchester to direct the Cavendish Laboratory at Cambridge University, where he had apprenticed under J. J. Thomson. The "old place," as he fondly viewed the Cavendish, was very congested now, with naval officers as well as researchers, and he meant to enlarge it.

Ellen sailed home in October 1920 to yet another accolade for her scientific achievements. The Fridtjof Nansen Prize, named for the great Norwegian polar explorer and oceanographer, was nearly as old as the Nobel Prize. The Nansen Trust presented the award annually in both the literary arts and the natural sciences, but it had never before recognized a woman as a winner in either category.

Distinguished and decorated, Ellen, now forty, resumed her teaching and research at the university. Her youngest sibling, Kristian, had moved to Trondheim to study civil engineering at the Norwegian

Institute of Technology, but her brother Adler, a maker of topographic maps, still lived with her.

In November, Ellen and her only female colleague on the faculty, Kristine Bonnevie, formed a local branch of a fledgling sisterhood known as the International Federation of University Women. The organization had been conceived just after the war, in the hope that educated women could help prevent another such catastrophe by nurturing global friendships. The IFUW particularly appealed to Ellen because it counted the creation of fellowships for study abroad among its main goals. "Women who have had such experiences will come home to their country with the most valuable of all gifts," she predicted, namely "a decision to continue their research."

Ellen's Curie-lab friend Eva Ramstedt agreed, and started a Swedish IFUW chapter in Stockholm. In Quebec, Harriet Brooks Pitcher took an immediate interest in the IFUW. Already active in the McGill Alumnae Association and a charter member of the Women's Canadian Club of Montreal—and having traveled twice to Europe on scholarships—she joined the federation's Canadian branch.

Harriet had last seen her mentor Ernest Rutherford in 1914, on his way back from a family reunion in New Zealand and the British Association meeting in Australia. On that occasion, Mary Rutherford and thirteen-year-old daughter Eileen had stayed at Harriet's home for several days while "Ern" tended to business at the University Club. Harriet still exchanged family-news notes with Mary, but the only letters she saw from Sir Ernest were the ones printed in the columns of newspapers and journals. On December 8, 1920, for example, in the midst of a Cambridge debate about granting women degrees and other privileges long reserved for men only, Rutherford and chemist William Pope aired their opinion in the *Times*.

"For our part," the men stated jointly, "we welcome the presence of women in our laboratories on the ground that residence in this University is intended to fit the rising generation to take its proper place in the outside world, where, to an ever increasing extent, men and women are being called upon to work harmoniously side by side in every department of human affairs."

In this climate of shared optimism and commitment to a peaceful future, Mme. Curie carefully considered the part that she would play. Numerous groups invited her participation or at least her signature on a declaration. Rather than ally herself with the IFUW in France, she agreed—along with Albert Einstein, Hendrik Lorentz, Kristine Bonnevie, and others—to be inducted into the International Committee on Intellectual Cooperation, an advisory arm of the new League of Nations. It was hoped that the brightest lights in the arts and sciences, by pooling their gifts, could envision a peaceful world and map a path to its realization.

Chapter Twenty

MISSY (Silver)

———

MARIE MATTINGLY MELONEY—known as "Missy" to her friends—needed nearly a year, and the help of an army of American volunteers, to raise $100,000. By late January 1921, she had the requisite sum in hand. For her next feat, she intended to import Mme. Curie and her daughters, and take them on a cross-country publicity tour highlighted by a stop at the White House to accept her gram of radium from the president of the United States.

Marie felt she could spare at most a couple of weeks for these diversions, but Missy Meloney wanted six weeks, or better yet, eight. Marie chose October as the ideal time to travel, after the summer vacation she needed and before the start of the academic year. Missy preferred May, when the visitor could be feted at the commencement exercises of select colleges and universities. Missy prevailed on both counts. She planned to sail to France in spring, then shepherd the Curies over the Atlantic and across America, accompanying them on all but the return leg of their two-month itinerary.

Marie, grateful for the great gift and the opportunity to thank her American benefactors in person, feared herself unequal to the physical and social demands of such a trip. The prospect of the numerous public appearances cowed her.

By now it seemed likely that a few of Marie's bodily ills, perhaps even her cataracts, were due at least in part to excessive radiation exposure. She had conducted her early exploits in the shed with scant concern for personal safety. The sight of the nighttime glow of radium exuded so much benevolent promise that she dismissed her injured fingers and the burn on Pierre's arm as manageable risks of exploring the unknown. Nor were the Curies alone in their disregard for radium's dangers. Ernest Rutherford once lost a tube of radium salt on the train between Montreal and Ottawa, and casually estimated that the tube would continue releasing radium emanation for several thousand years. In 1921, however, both the *Institut du Radium* and the Cavendish Laboratory initiated programs of regular blood testing. When lab members' red or white blood cell counts dipped below normal, they were advised to take a few days off to "rebuild" them.

Marie saw Sir Ernest again the first week of April, in Brussels, when the third Solvay Council convened to ponder "Atoms and Electrons." A young Danish physicist named Niels Bohr was expected to attend the conference to expound his theory of the way an atom's electrons distributed themselves around the nucleus in a series of nested orbital shells. But unfortunately Bohr pled illness from overwork and stayed home in Copenhagen. Albert Einstein, the only German-born scientist asked to enter Belgium so soon after the war, declined the invitation and sent a written contribution. Two Americans, Albert Michelson and Robert Millikan, were the first of their countrymen to join the exclusive gathering. Once again, Mme. Curie stood out as the only woman among the twenty-five participants.

"I'm not doing too badly, but I am tired," she wrote her daughters midway through the week. The discussions went on all day, she said, and the topics were so interesting that one wore oneself out taking everything in. "With meetings like this, I could visit Brussels a hundred times and never know the city any better than I did the day I was born." She hoped that her stay in America would allow her to see some of that nation's storied natural wonders.

From her *Olympic* cabin in mid-Atlantic on May 10, she wrote her friend Henriette Perrin, "I found your sweet letter on board, and it did me good, for it is not without apprehension that I have left France to go on this distant frolic, so little suited to my taste and habits." Indeed, even before she stepped ashore in New York, a crowd of thousands had assembled at the pier to welcome her with music, flags, flowers, and other signs of adulation.

The moment proved an awakening for Irène and Ève. Upon their arrival in New York, as Ève recalled in her biography of her mother, the two sisters "discovered all at once what the retiring woman with whom they had always lived meant to the world." Ève herself, now sixteen and easily the most attractive and fashionable member of the family, earned a press epithet as the girl with "the radium eyes."

Missy Meloney, Irène, Marie, and Ève Curie arriving in New York

The Curies' first sightseeing excursions took them to several women's colleges in bucolic settings—Vassar, Smith, Mount Holyoke. Marie marveled at the attention these institutions lavished on exercise and well-being. "A very complete organization of games and sports exists in every college," she observed. "The students play tennis and baseball; they have gymnasium, canoeing, swimming, and horseback riding. Their health is under the constant care of medical advisers." At the Sorbonne, in contrast, the recently opened sports park was limited to men only. "Don't our daughters need exercise and good health?" Professor Curie had demanded in a formal complaint to the rector. Now she could not help but notice "important differences between the French and American conception of girls' education, and some of these differences would not be in our country's favor."

Ève, soon to begin her own higher studies, said she saw "white-robed girls in line along the sunny roads; girls running by the thousand across grassy slopes to meet Mme. Curie's carriage; girls waving flags and flowers, girls on parade, cheering, singing in chorus . . . Such was the dazzling vision of the first days."

Administrators and alumnae of the women's colleges were major contributors to Missy Meloney's "Marie Curie Radium Fund," and many of them numbered among the more than three thousand attendees at the tribute to Madame in New York's Carnegie Hall on May 18. The evening focused on women in science, with guest speakers including Dr. Florence R. Sabin of the Johns Hopkins Medical School, and a presentation of the Ellen Richards Research Prize of $2,000 to Mme. Curie. Another New York event in her honor drew several hundred representatives of American scientific societies to a luncheon at the Waldorf Astoria.

Next the entourage moved on to Washington, DC, for the official presentation ceremony. Barely a week into her US visit, Marie had already been embraced and congratulated so repeatedly and enthusiastically that her right wrist was sprained and she needed to keep her arm in a sling. On May 20, in the Blue Room of the White House, before secretaries of the cabinet, justices of the Supreme

Mme. Curie with President Warren Harding

Court, foreign diplomats, and high-ranking officers of the army and navy, President Warren Harding made a speech affirming the historic bonds of friendship between the peoples of Poland, France, and the United States.

"The radium itself was not brought to the ceremony," Marie noted in her account of the event. "The President presented me with the symbol of the gift, a small golden key opening the casket devised for the transportation of the radium." This turned out to be a mahogany treasure chest, about one foot square, weighing more than one hundred pounds (forty-six kilograms) because of its lead lining, and custom-fitted to nestle ten tiny glass tubes containing a gram's worth of radium.

America's gift of radium in its
lead-lined box

The precious content remained for the time being in safekeeping at the National Bureau of Standards, where Marie got a glimpse of it two days later. She understood —better than anyone—the need to certify her radium against the American secondary standard held at the bureau. Besides, she could not conveniently tote the material from place to place along her route.

The gift of her radium came from carnotite ore mined in Colorado and processed at the Standard Chemical Company's plant in Canonsburg, Pennsylvania. When Marie toured the facility with the company president and plant manager, she learned that carnotite was a much poorer source of radium than the St. Joachimsthal pitchblende she had sifted en route to her great discovery. And yet, she was pleased to point out, "the means of extraction of radium are still the same" as the ones she originated and explicated in the text of her 1903 dissertation.

At Canonsburg, as at the Bureau of Standards and the Laboratory of Mines, she conversed energetically with employees who pledged their allegiance to scientific research. What drained her were the rounds of banquets and pageantry. Her entry into each new city stirred up more excitement than she could withstand, even before the scheduled formalities began. She was repeatedly asked to don academic robes to receive an honorary degree, either Doctor of Law or Doctor of Science, from the Universities of Pennsylvania and Chicago, Pittsburgh University, Northwestern, Yale, Columbia, Smith College, Wellesley College, and the Women's Medical College of Pennsylvania. Marie found it impossible to sit through so many commencement exercises and accept every diploma, so sometimes Irène or Ève stood in for her. And still Marie could not get enough rest to counter her fatigue. At length Missy canceled half the engagements and gave her guests time off to cross the country by train and gape at the Grand Canyon. While Irène and Ève and other tourists rode mules to the bottom of the world's widest, deepest river gorge, Marie steeled herself for the final few obligations back East. She collected enough mental pictures "of the great falls of Niagara and of the magnificent colors of the Grand Canyon" to keep both vivid in her mind.

In a lengthy letter to Ernest Rutherford, Bertram Boltwood reported on his meetings with Marie in the United States.

> I saw the Madame first at a luncheon given in her honor in New York shortly after she landed. Then I saw her again at New Haven when she came for Commencement. Kovarik and I had her for a couple of hours at the Sloane Laboratory and I was quite pleasantly surprised to find that she was quite keen about scientific matters and in an unusually amiable mood, although she is in very poor physical condition and was on the verge of a breakdown all the time she was over here. She has learned a lot of English since we saw her in Brussels [in 1910] and gets along quite well in a conversation. She certainly made a good clean up over here and took back a gram of radium and quite a tidy number of thousands of dollars. But I felt sorry for the poor old girl, she was a distinctly pathetic figure. She was very modest and unassuming, and she seemed frightened at all the fuss the people made over her.

On June 28 she and her daughters reboarded the *Olympic*, homeward bound. The box of radium, delivered dockside, found a berth in the purser's safe for the weeklong voyage. No throngs of admirers crowded the port at Cherbourg to greet the ship, and the welcoming committee in Paris at the Gare Saint-Lazare consisted solely of Marcel Laporte, a rising physicist from the *Laboratoire Curie*. Though the hour was late, Laporte took the radium directly to Madame's office at the institute.

Throughout their travels together in America, Missy had pressured Marie to write her autobiography. This was an alien concept to Marie, who protested that her life story could be summed up in a single paragraph, and she dictated it flatly to prove her point: "I was born in Warsaw of a family of teachers. I married Pierre Curie and had two children. I have done my work in France." She still believed what she had told the reporters who invaded the shed after the 1903 Nobel Prize: "In science, we should be interested in phenomena, not in individuals." Missy, equally tenacious, persevered and also contacted several publishers she knew. The Macmillan Company offered the reluctant author 50,000 francs as an advance against royalties. Marie could not refuse such a bounty for her laboratory, desperate though she was to step out of the spotlight. She compromised by consenting to write a biography of Pierre, with just a brief sketch of her own life tacked on as an appendix.

But first she needed to finish her work in progress—a handbook of about 150 pages, aimed at laboratory researchers and summarizing the current knowledge of isotopes. She worked on the manuscript that summer in Paris, while her daughters swam at l'Arcouest, and proofread it at Cavalaire during a working holiday in September. The Society of Physics published *L'isotopie et les éléments isotopes* before the close of the year. Another, even shorter book she wrote about her experiences as a radiologist during wartime, *La radiologie et la guerre*, was also published in 1921, by the firm of Félix Alcan.

"It is not without hesitation," Marie allowed when she approached her most challenging assignment, "that I have undertaken to write the biography of Pierre Curie. I should have preferred consigning this task to some relative or friend of his infancy who had followed his whole life intimately and possessed as full a knowledge of his earliest years as of those after his marriage." Pierre's brother, Jacques, helped her along by generously sharing his memories, as did Pierre's friends, and of course Pierre himself had told her many tales of his youth. Thus armed, she recreated Pierre's family of origin, his progress from childhood to first scientific work, the discovery of piezoelectricity, and his early experience as the director of student laboratories in the

School of Physics and Chemistry, where he found time and space for his own studies of symmetry and magnetism. She needed no assistance to recall or express "the profound impression his personality made upon me during the years of our life together."

While she could not build a complete narrative, and certainly not a perfect one, she hoped that her portrait of Pierre would "conserve his memory" and "remind those who knew him of the reasons why they loved him."

Chapter Twenty-One

CATHERINE
(Mesothorium)

———

ADOPTED AS AN American heroine, Marie rose to new stature in France as well. Toward the end of 1921 her friend Dr. Antoine Béclère, the eminent radiologist, circulated a petition among his fellow medical men in support of her election to the *Académie de Médecine*, and on February 7, 1922, a large majority of them voted her in.

"We salute in you a great scientist," said academy president Anatole Chauffard by way of welcome, "a great-hearted woman who has lived only through devotion to work and scientific abnegation, a patriot who, in war as in peace, has always done more than her duty. Your presence here brings us the moral benefit of your example and the glory of your name."

Had she been inducted a decade earlier into the *Académie des Sciences*, she would have breached the all-male bastion of the *Institut de France*. Within the free-standing medical academy, she became an even stranger anomaly—not only the first woman but also the sole physicist among 110 physicians. Her inclusion underscored the increasingly vital role of physics in medicine. Marie's own Curie Foundation had begun building a pair of new dispensaries, one for radium therapy and one for X-ray diagnosis and treatment, just around the corner from the Radium Institute on the rue d'Ulm.

"You are the first woman of France to enter an academy," Dr. Chauffard reminded the honoree, "but what other woman could have been so worthy?"

Marie, who was not seeking uniqueness, continued admitting women to the *Laboratoire Curie*. The year 1921 had brought two more. Jeanne Lattès, a thirty-three-year-old widow and mother, was pursuing her doctorate in physics at the Sorbonne and cancer research with a colleague in the *Pavillon Pasteur*. Catherine Chamié arrived with her physics doctoral degree already in hand, but feared she no longer possessed the capabilities she once had mastered. As she confessed in her first appeal to the director, "I will soon lose the moral right to use my title as *docteur es sciences*."

Catherine Chamié, born in Odessa in 1888 to a Russian mother and Syrian father, studied physics at the University of Geneva. She joined a research team in Petrograd, but with the outbreak of the Great War she turned her talents to nursing. No sooner had she resumed her scientific activities than the 1919 civil war in Russia forced her to flee the country with her widowed mother and siblings. They spent five months in a refugee camp near Grenoble before reaching Paris, where, of new necessity, Catherine became a teacher at the *Lycée russe*. She was still employed at the Russian secondary school when she approached Mme. Curie for a position, and she held on to her teaching job as she began spending her spare hours in the lab as an unsalaried independent researcher.

It felt good to handle real scientific instruments again, and to learn new skills for carrying out experiments in radioactivity. "The series of operations," Catherine noted while observing the master at work, "was effected by Mme. Curie with admirable discipline and harmony of movement. No pianist could have performed with greater virtuosity what her hands accomplished. It was a perfect technique, which tended to reduce the coefficient of personal error to zero." She could only hope her own efforts, so clumsy in comparison, would improve with practice.

Some months later a letter of commendation from the director of the *Lycée russe* finally reached the director of the *Laboratoire Curie*. It said Mlle. Chamié deserved the highest praise for the depth of her knowledge, her zeal, and her conception of duty. By this time Marie could see as much for herself, and she arranged for the emigrée to receive a Curie-Carnegie fellowship of 4,000 francs for the 1921–1922 academic year.

Marie paired her with physicist Dragolijub Yovanovitch, a native of Belgrade, and urged them to improve the methods for preparing and preserving homogeneous solutions of radium salts. All too often, such solutions developed solid clumps that clung to the walls of their containers, trapping emanation and thereby confounding the measurement of emanation release. The new lab partners worked together on this problem every afternoon, and Catherine often spent the evening as well, while continuing to teach morning classes at the *Lycée russe*. She had no choice, as she needed both incomes to support her family. Fortunately, the experiments boded well and she secured the promise of a second Curie-Carnegie fellowship for the coming year.

———

JULY 1922 FOUND Marie in Cavalaire for a brief rest before the daunting first meeting of the International Committee on Intellectual Cooperation, to be held in Geneva in early August. As often happened when she traveled, she caught a cold. Between her illness and the enervating effects of the southern wind—the mistral—she could get hardly any writing done, and had not even glanced at the dossiers provided by the League of Nations. At least, she reported to her daughters in l'Arcouest, the mistral had completely driven off the mosquitoes.

Marie anticipated seeing Albert Einstein again in Geneva, as he, too, had been named to the new committee of intellectuals. "I believe your acceptance, as well as mine," she had written him then, "is necessary if we have any hope of rendering any real service," and he had agreed with her sense that "the League of Nations, although still imperfect, is a hope for the future." In late June 1922, however, in the wake of the shocking assassination in Berlin of Jewish industrialist and foreign minister Walter Rathenau by paramilitary political rivals,

Einstein changed his mind. Aware that "a very strong anti-Semitism reigns in the milieu that I am supposed to represent to the League of Nations," he wrote, he felt strongly that "a Jew is not the best person to serve as a liaison between the German intelligentsia and the international intelligentsia." Moreover, the League of Nations, established in Paris at the end of the Great War, did not yet count Germany among its member states.

In Geneva, as in Brussels, Marie's time was consumed by her participation in the committee meetings. "You will be astonished to learn that your mother is not content to remain mute," she reported home, "but contributes frequently, evidently deluding herself that doing so will be of use!" Her fellow members included Paul Langevin, Hendrik Lorentz, Robert Millikan, and other men she had engaged with repeatedly at scientific conferences, along with prominent figures in the liberal arts. But in Geneva, unlike the usual Brussels situation, Marie was no longer the only woman. Biologist Kristine Bonnevie of Norway, the first female professor in her native country and the first woman elected to the Norwegian Academy of Science and Letters, also gave voice to her opinions at the lakefront *Palais des Nations*. Professor Bonnevie directed the Institute for Inheritance Research at the Royal Frederick University in Kristiania. Never married, nearing age fifty, and the veteran of political office in both the Kristiania city council and the national parliament, she had recently agreed to serve as president of the Norwegian branch of the International Federation of University Women.

Marie emerged from the Geneva deliberations with membership in two subcommittees, one devoted to securing scholarships and grants for researchers, and the other to creating an annotated—and constantly updated—bibliography of scientific articles from publications the world over. If she had wondered, before the gathering, what the commission might accomplish and how she could contribute, she now had a sense of purpose.

At her official post in Paris, she interacted with her students much more than she had in the past at the Annex. The increased contact owed a good deal to the architecture of the Radium Institute, which consolidated the scattered enclaves of the old Curie lab. And whereas she had formerly left most student supervision to André Debierne, the *chef de travaux*, she no longer relied on him to fulfill the mentor's role. His experiences during the war had profoundly affected him. The man Sybil Leslie once described as remarkably "gentle, kind, and courteous," and possessing "a vast fund of patience," could now be heard grumbling his comments to students, and sometimes outright growling at them. Madame, ever more confident in conversation, scheduled student consultations as needed, joined casual discussions in the hallways, and stopped on the steps to chat with one or another on the way up to her office at the start of work each morning. Often these impromptu encounters drew in other staff members and visiting researchers for sessions that the students came to call "the study of the science of radioactivity on the staircase."

If Marie felt any lingering sting from her 1911 rejection by the *Académie des Sciences*, she no longer shunned its prestigious weekly publication, the *Comptes rendus*. In 1921, for the first time in a decade, she had submitted a report of her recent research, to be read into the record by her former professor Gabriel Lippmann. In it she acknowledged the difficulty of separating radium from its isotope mesothorium, which was used to make luminous paint, and she offered a means for deriving accurate radioactivity measurements of each substance in any given commercial sample.

After Lippmann's death in the summer of 1921, Marie turned to other *Académicien* friends who were just as willing as he had been to represent the members of her lab. Chemist Georges Urbain, for example, read the 1922 joint report by Dragolijub Yovanovitch and Catherine Chamié, describing the effective stirring apparatus they had assembled for keeping radium solutions from clumping.

The partnership of Mlle. Chamié from Russia with the Yugoslavian M. Yovanovitch typified the international makeup that had long characterized the Curie lab. In 1922 the roster also listed Stefania Maracineanu from Romania and radiochemist Sonia Slobodkine from Poland. Another Pole, Salomon Rosenblum, came in 1923, after studying under Niels Bohr in Copenhagen. Nobuo Yamada arrived at the same time from Tokyo Imperial University, having been selected by the Japanese government to become his country's first radioactivist.

The Curie lab, a tightknit marital unit in its formation at the turn of the century, remained a family enterprise in the 1920s, with a mother-daughter team at its heart, and nepotism personified by Madame's nephew Maurice. Although Ève Curie had avoided involvement in radioactivity, she was well known to the lab personnel, especially to her "Uncle André," who had recently coached her through the arduous preparation for her *baccalauréat* exams. The other scientists were pleased to receive tickets to her piano recitals. She was hoping to play again soon for Ignacy Paderewski.

Marie had wanted Ève to attend medical school and pursue a career as a specialist in the treatment of cancer by radioactivity. But rather than try to sway her younger daughter's choice of profession, she indulged Ève's obvious talent by buying her a grand piano. It was the only extravagance to be found in their home.

Irène and Marie Curie, 1922

"That apartment on the quai de Béthune," Ève later wrote, was "very large and not very comfortable," making for "a strange sort of family dwelling." It occupied the third floor of a building that dated from the time of Louis XIV, and its rooms seemed, to Ève at least, to cry out "for the majestic armchairs and sofas that would have suited their proportions and their style." In place of such period décor stood the old mahogany parlor suite inherited from Dr. Curie, with its pieces "grouped at random in the huge drawing room—which was big enough for fifty but rarely held more than four." No carpets hid "the skating rink of a fine waxed parquet that creaked and complained under one's feet." No curtains blocked the view of the Seine through the tall windows.

"The only room in the house that produced the emotion of life," Ève judged, was her mother's study. "A portrait of Pierre Curie, glassed-in shelves of scientific books, and a few pieces of old furniture created an atmosphere of nobility there."

MARIE WORE EYEGLASSES now, at age fifty-five, and also needed to place colored marks on the dials of her meters in order to read them. Her lecture notes were writ large. When students showed her photographs of spectral lines, she only pretended to assess the images, and actually extracted the information by asking astute questions. Her associates noticed the ruse, though they pretended not to. "Perhaps radium has something to do with these troubles," Marie suggested in a letter to her sister Bronya, "but it cannot be affirmed with certainty."

In mid-July 1923, Marie asked Ève to leave l'Arcouest and stay with her in the city for a few days while she underwent cataract surgery on both eyes. "Tell our friends that I have been unable to complete a piece of editing that we were working on together, and that I require your help to finish it quickly." No one in the Brittany circle needed to know the real reason for Ève's sudden return to Paris. A postscript reiterated Marie's desire for secrecy: "Tell them as little as possible, *chérie*."

After the operations, Irène took over as her mother's aide during a recuperation retreat at Cavalaire. "I am getting used to going without glasses and have made some progress," Marie wrote to Ève in late September. "I joined two walks: one afternoon excursion and one day-long, over some rough mountain trails. They went off well, and I can move quickly enough without accidents." The most bothersome aftereffect was a persistent double vision that kept her from recognizing people as they approached her.

"Every day I do some exercises in reading and writing," she added, and these proved more strenuous than hiking.

At the end of December 1923, Marie appeared once again in the public eye at the Curie Foundation's gala observation of the twenty-fifth anniversary of the discovery of radium. Bronya, Józef, and Helena traveled together from Warsaw for the event. Around them in the great amphitheater of the Sorbonne sat delegates from French and foreign universities, scientific societies, student associations, and government offices. The night's entertainment featured a staged reading, by André Debierne, of the Curies' original discovery paper that Henri Becquerel presented to the *Académie des Sciences* on December 26, 1898. "On a New and Strongly Radio-active Substance Contained in Pitchblende" had lost none of its crispness in the lapse of years.

A lively demonstration of radium experiments followed, performed by Irène Curie and Fernand Holweck. Then the president of the republic, Alexandre Millerand, offered Mme. Curie France's gift of a "national recompense"—an annual pension of 40,000 francs, approved by unanimous vote in Parliament—"as a feeble but sincere witness of the universal sentiment of enthusiasm, respect, and gratitude which follow upon her."

Chapter Twenty-Two

FRÉDÉRIC (Radon)

SINCE THE RUPTURE of the "Langevin affair" of 1911, Marie and Paul had strengthened the ties of friendship, common interests, and shared ideals that drew them together in the first place. They saw each other often, whether casually through social connections or formally via the Solvay Council, the International Committee on Intellectual Cooperation, and assorted Parisian scientific meetings. In November 1924, when Paul recommended one of his former students for a position at Marie's lab, she agreed to see the young man immediately, even though he lacked experience in radioactivity.

Frédéric Joliot had studied physics with Langevin at Pierre's old school, the *École Municipale de Physique et de Chimie Industrielles*, then worked briefly as an engineer at a steel plant in Luxembourg. Now twenty-four and nearing the end of his military service, he showed up in uniform for his anxious initial interview with the director. Since she asked only one question—When could he start work?—he told her nothing of his childhood exploits as an avid experimenter, or that, as a boy of six, he had clipped a photograph of Pierre and Marie Curie from a popular magazine, and still kept it hanging, framed, in his bedroom.

Joliot returned to the Radium Institute in mid-December, in civilian clothes. Madame assigned him to assist René Cailliet, a *préparateur* on the verge of retirement. She also counseled him to pursue his higher education while acquiring new lab skills. She held up her daughter as an example: Irène was only three years older than he, and already at work on her doctoral dissertation.

Irène's thesis project concerned polonium, the first of her parents' two element discoveries. Specifically, she sought to study its alpha radiation—the positively charged particles that polonium emitted as it decayed. Aware that scientists in several countries were similarly engaged, Irène did not let concern for their potential competition deter her. She planned to clock the initial speed of the alphas and the distances they covered—in centimeters—under various conditions. She would also track their ability to ionize the gases in their path as a function of distance traveled.

In January 1925, only a month into Frédéric Joliot's tenure, bad news reached the Curie lab about two of its former students. Marcel Demelander and Maurice Demenitroux, working together at a small factory near Paris that produced thorium and mesothorium for medical use, had died within days of each other: first Demelander, age thirty-five, of a rapid-onset anemia, then Demenitroux, forty-one, of leukemia. Mme. Curie learned from their colleagues that both had been confined to cramped, poorly ventilated spaces, and denied the proven safeguards of lead screens, periodic blood tests, and frequent breaks for fresh air.

In response to these fatalities, the French Academy of Medicine appointed its newest member, Marie Curie, to a study commission that included Drs. Antoine Béclère and Claudius Regaud.

Their review of industrial working conditions revealed a widespread and alarming disregard for the safety practices observed at research laboratories. Those established precautions, the commissioners insisted in their report, must be widely adopted—and adapted to the specific situations and techniques in use in each venue. Any facility that handled or even transported "radioactive bodies," they recommended, should be classified as "insalubrious," and therefore subject to regulation by the Ministry of Work and Health.

Beyond her official participation in the matter, Marie took up a collection to aid the widows Demelander and Demenitroux.

Surely, she believed, the stated guidelines would suffice as protection. After all, no one at the *Institut du Radium* suffered any impairment. She herself had doubtless logged more lifetime exposure to radium than anyone in the world, yet here she was, a woman of slight build and nearly sixty years, still coming to the lab every day, still actively engaged in productive research. She did wonder sometimes, in her letters to Bronya, whether this or that bodily complaint could be blamed on radium, but only her hands showed signs of damage, and those burns had occurred long ago. Although she had likely inhaled more than her fair share of radioactive gas, her intake had obviously not amounted to a debilitating dose, let alone a lethal one.

Radium emanation had at last gained an official name, "radon," approved by the recently established International Union of Pure and Applied Chemistry. "Radon" neatly abbreviated the "radioneon" suggested by Mme. Curie, and also achieved a pleasant harmony with the other noble gas names—argon, krypton, neon, and xenon. Many radioactivists referred to thorium emanation as "thoron," and actinium emanation as "actinon," though both of these were understood to be isotopes of radon, element number 86 on the periodic table. Harriet Brooks, who once hesitated to call the "new gas from radium" an element, and then spent years investigating the three emanations at laboratories in three different countries, had a third child now, Paul Brooks Pitcher, and a reputation in Montreal as an excellent gardener.

———

IF MARIE HAD been ignorant of risk and heedless of danger in her youth, she had nevertheless lived to tell about it, and was now doing everything in her power to assure the proper handling of radioelements in research and industry. She judged the Curie lab environment safe enough to serve as her own daughter's workplace. In fact, from a coldly analytical standpoint, her two children constituted a natural experiment in radioactivity exposure. Irène had been a fixture in the Radium Institute for five years now, while Ève rarely entered the building. Between them, Irène boasted the more robust constitution.

Ève had just completed her formal education at the *Collège Sévigné* —the same school Irène attended after the co-op experiment ended— and was pursuing private music studies with pianist Alfred Cortot. The sisters remained as close as could be expected, given the differences in their ages and tastes. They even shared a few childhood friends among the regulars at l'Arcouest.

Ève always maintained that Marie treated her daughters equally, showing no favoritism. Her mother's only fault, according to Ève, was to indulge the artistic teenager's whims, letting her choose her own music teachers and methods of work, and enthusiastically encouraging her in one "capricious" plan after another. "She was bestowing too much freedom upon a being undermined by doubt," Ève judged in retrospect, "who would have done better to obey firm indications." At the same time, Ève understood her mother's mindset—"she who had been led to her destiny" by "infallible instinct" and against "immense obstacles." Ève saw the same drive, the same sort of innate resolve, in her older sister, their mother's clone. With a mix of humor and bravado, Ève made light of her exclusion from the scientific discussions that dominated household conversation. As a little girl, she said, when she heard the two of them exchange algebraic expressions such as "Bb squared" and "BB prime," she had pictured *bébés* like herself, but oddly shaped to fit in corners, or given special privileges.

On March 27, 1925, Irène upheld the family tradition by successfully defending her doctoral research at the Sorbonne. Because of her family history, the three professors who examined her could not help but be family friends of long standing, whom she had trusted as father figures all her life. André Debierne, Jean Perrin, and Georges Urbain deemed Mlle. Curie's polonium work deserving, and the university awarded her the *docteur ès sciences* degree.

Frédéric Joliot, the recent hire, attended Irène's thesis defense and congratulated her on the spot. Later that day, in the garden of the *Institut du Radium*, Paul Langevin joined the traditional celebration that Madame held to honor a new *docteur* at the Curie lab. Everyone drank tea out of beakers, with pipettes for stirrers, and ate cakes and cookies served in basins borrowed from the photography darkroom.

Now Irène had more time to coach Joliot, whose further formation as a radioactivist had fallen to her. He obeyed her every instruction. He also did as Madame had advised him, and in July he passed his second *baccalauréat* with distinction. By then he had befriended the other researchers, students, and staff, a few of whom were amused by the almost comical mismatch between tall, taciturn Irène and her handsome, gregarious acolyte.

Frédéric Joliot

As Frédéric himself later said of Irène, "With her cold exterior, forgetting sometimes to say good morning, she did not arouse a feeling of sympathy in the lab. But I discovered in this young woman, whom the others saw somewhat as a block of ice, an extraordinarily sensitive and poetic person who in many ways was the embodiment of what her father had been. I had read much about Pierre Curie, I had heard from teachers who knew him, and I found in his daughter the same simplicity, good sense, and humility."

ONE DAY Catherine Chamié arrived from her morning of teaching at the *Lycée russe* and asked Mme. Curie to help her with an experiment. The lengthy process stretched through the afternoon and evening hours to well past midnight. On such occasions, when Madame did not leave the lab with her daughter as usual at the end of the day, Frédéric walked Irène home. And when the director left Paris in June for several weeks of travel, Frédéric became Irène's more constant companion on long rambles through the forest of Fontainebleau.

"I find myself altogether too far away from you," Marie wrote home to her daughters on June 3, 1925, from Warsaw, where she had gone to mark the start of construction on a new radioactivity research and treatment center. The city's first radium institute, established in 1913, had perished with its on-site director, Jean Kazimierz Danysz, in the early months of the Great War. In recent years Bronya had raised funds to rebuild it by sending out thousands of postcards exhorting the populace to "buy a brick for the Marie Sklodowska-Curie Institute!" The same cards carried a quote from Marie, saying, "My most ardent desire is the creation of an institute of radium in Warsaw."

At a sunny morning ceremony, the president of the republic, Stanislaw Wojciechowski, laid the first stone, Marie the second one, and the mayor of Warsaw, Wladislav Jablonski, the third. Other fetes followed for her at universities and learned academies. One orator hailed her as "the first lady-in-waiting of our gracious sovereign, the Polish Republic." To some people, she was the very personifi-cation of radium.

After Poland, Prague. "I am astounded by the life I'm leading," she wrote Irène on June 14, "and incapable of saying anything intel-ligent at this moment." The next day would take her to Jáchymov (formerly Joachimsthal), to the mouth of the uranium mine that had yielded her original store of pitchblende. The life she was leading on this trip retraced the course of her old life, returning her as a conquering heroine to places where she had once been compelled to study in secret or to beg for mine tailings discarded as waste. Yet she felt depleted by all the fuss being made on her behalf. "I ask myself, what fundamental flaw of human nature makes people believe this form of agitation is necessary?" At the same time, she did not doubt the sincerity of her hosts in their determination to do every thoughtful thing for her. "I am here in a magnificent apartment, bedroom, salon and bath, with a view of the river and the hills beyond, surrounded by flowers given to me at the station, mostly roses because this is their season."

———

MARIE RETURNED FROM the exertion and adulation of her tour to confront a new revelation about the marvel that was radium. A letter from Missy Meloney in America said that several young women whose job was to paint the numerals on luminous watch dials—with radium-laced pigment—had died of what appeared to be an occupational illness. Their employer, a private plant called the U.S. Radium Corporation, had opened in 1917 in Orange, New Jersey, to make glow-in-the-dark wristwatches for soldiers in the trenches, and now sold them to civilians. One currently afflicted dial painter, Margaret Carlough, was suing the company. Too disabled to remain at her job, she claimed that the recommended practice of pressing the fine tip of the dipped paintbrush between her lips to achieve a perfect point had destroyed her jaw.

"Lip-pointing" caused a dial painter to ingest a little of the paint, and with it a taste of radium, which was so like calcium in its chemistry that it readily settled into teeth and bones. Once lodged there, radium's characteristic radiations—and those of its decay products— pervaded the body and destroyed the bodily tissues. When doctors tested Margaret Carlough's breath during a diagnostic examination, they found that she exhaled radon along with the ordinary products of respiration.

More disturbing news reached Marie from a chemist she had met in America—Harlan Miner of the Welsbach Company in Gloucester, New Jersey, makers of mantles for gas lanterns. Miner was mourning "the quite sudden death from anemia of one of the young men in our organization," and reported "the almost simultaneous death from the same disease" of another scientist at a nearby radium manufacturing company. Regular blood testing did take place at Welsbach, Miner's letter said, but had apparently been instituted too late. "The man in our employ who died from anemia had a very low number of red corpuscles when we finally came to have him examined; and although he received blood transfusions it was not possible to save his life."

Marie had never advocated the use of radium in commercial paint—or for any application outside a laboratory or hospital setting. The addition of thorium to the fabric of gas mantles had been common practice long before she recognized thorium's radioactivity. And yet, responsibility for a host of ills seemed to be settling now on her shoulders.

Part Four

Large-Scale Production Facility

Arcueil

Many imagine it was like a jigsaw—
but no, it was more like cataloging
a collection of rare, exotic birds:
examining how the plumage
of beryllium resembled magnesium,
the curve of the wing,
the clutch of their feet
on prey.

—Jennifer Gresham,
"Building the Periodic Table,"
Diary of a Cell

Today's ceremony is a deliberate outreach on our part from the *Panthéon* to the first lady of our honored history. It is another symbol that captures the attention of our nation and the exemplary struggle of a woman who decided to impose her abilities in a society where abilities, intellectual exploration, and public responsibility were reserved for men.

—French president François Mitterrand, speech of April 20, 1995, when Marie Curie's remains were interred at the *Panthéon*.

Chapter Twenty-Three

ALICJA (Polonium)

———

IRÈNE WAS IN LOVE. One morning at breakfast with her mother and sister, early in 1926, she announced her intention to marry Frédéric Joliot. She believed the harmony they had achieved as lab partners, despite their different temperaments, promised equal compatibility as husband and wife. Her memories of her own parents' union offered the perfect model for her mental picture of married life.

Marie, who relished the constant companionship of her daughters, tried to put the best face on Irène's decision. At least the impending marriage would not carry her off to some distant country, but only as far as another neighborhood of Paris. That was a comfort. Marie would still see her in the lab almost every day. And nothing needed to change immediately. For now, Frédéric remained Irène's subordinate, still working toward his doctoral degree. Anything might happen to alter the course of events.

"*Chère Mé*," the postcard from the ski resort at Megève began, dated February 19, 1926, in the not yet familiar hand of Frédéric Joliot. "Irène and Ève are in perfect health," he beamed, "and comport themselves valiantly over the frozen snow. Ève in particular wins admiration for her first attempts. We are gathering our strength and thinking often of the laboratory and the work that awaits our return. We send you our love." It was signed "Fred" and Irène.

The pace of the engagement had accelerated. Marie voiced no objections, though according to Ève, she "tried in vain to conceal her inner dismay."

While Irène was away, Marie walked home from the lab on the arm of Alicja Dorabialska, a new independent researcher she had personally recruited. Alicja was the same age as Irène but had grown up more like Marie, in a proudly Polish family chafing under Russian rule in Sosnowiec. As a child, Alicja had heard tales of Mme. Sklodowska Curie that inspired her to become a scientist herself. Fortunately, times had changed enough in Poland for Alicja to earn an advanced degree in organic chemistry from the Warsaw University of Technology. She had been living in Warsaw, working at the Institute of Chemistry, when Madame paid a state visit in 1923 to mark the twenty-fifth anniversary of the discovery of polonium and radium. Their chance meeting precipitated an invitation to the Radium Institute, and now their evening strolls from the rue Pierre-Curie to the Île Saint-Louis gave them the privacy to converse in dialect.

At the lab Alicja collaborated with Dragolijub Yovanovitch in observations of the energy release from chemical reactions involving radioelements. Marie was writing a detailed summary for a Polish scientific journal about the chemistry of polonium—everything from the earliest investigations of the element's behavior to the methods for obtaining it, the results of recent experiments conducted at the Curie lab, and her interpretation of those results.

One could still extract polonium from mineral ore, as she and Pierre had done early on, but nowadays it was possible to mine polonium from used ampoules of therapeutic radon. Doctors and hospitals had no further use for these items once the encapsulated radon gas had exhausted itself in treating patients' tumors. Although the radon was gone, along with its short-lived daughters, a valuable residue of long-lasting breakdown products survived inside the ampoule. These included radium F, otherwise known as polonium, and an accumulation of radium D, a long-lived radioactive isotope of lead that would go on generating polonium for years to come.

Radon's decay gave rise to polonium three successive times in the cascade of transmutations that ended at ordinary—or nonradioactive—lead. Short-lived radium A showed up first, followed several stages later by the even shorter-lived radium C-prime, and finally the relatively stable radium F, with its 140-day half-life. Repeat appearances of the same element in the family genealogy illustrated the "laws of displacement" governing all radioactive decay. Every transformation "displaced" an element from one box to another on the checkerboard grid of the periodic table. When an atom of radium, for example, released an alpha particle, it lost two positive charges and transmuted to radon, two boxes to the left. Radon, too, decayed by alpha emission, to become polonium, another two boxes to the left. But beta emission worked in reverse: An atom that released a beta particle lost a negative charge of minus-one, or, said another way, it gained one unit of positive charge, and moved one box to the right. Thus, the direction and magnitude of change depended on the type of particle emitted at each stage. It was still not clear to anyone how a negatively charged beta particle could emerge from the hotbed of positivity in the atomic nucleus, but somehow it did, and several physicists had tendered their explanations.

The combination of alpha- and beta-type transformations caused the line of descent from atomic number 92 (uranium) to atomic number 82 (lead) to zigzag back and forth instead of following a straight path. As in the expression "two steps forward, one step back," two successive beta emissions followed by one alpha emission would render an atom an isotope of its ancestor, landing it back in its grandparent's place on the periodic table.

The serial displacements of the uranium family coexisted with those of the actinium family and the thorium family, creating a bustle of to-and-fro activity across the board in the uranium-to-lead zone. Strangely, none of the progeny ever settled in the still-empty space at number 87 on the periodic table, between radium and radon. Nor was there an occupant for the number 85 place, between radon and polonium. Perhaps these two unknown elements—sure to be radioelements—had already manifested at the *Laboratoire Curie* but passed through so quickly, on account of ultra-brief half-lives, that no one had noticed them.

IN JUNE 1926, after Frédéric discussed the wedding plans with his mother, Marie asked Emilie Roederer Joliot to lunch with her at home. They spoke of their children—Mme. Joliot had borne six but lost two in infancy and a third son in the war—all afternoon. Each widow made a good impression on the other, as expected. On the twentieth, Irène lunched with Mme. Joliot, who pronounced her "charming with us and affectionate with Fred." Ève met Frédéric's sisters, Jeanne and Marguerite, at a family dinner held in the Joliot apartment on the avenue d'Orléans to celebrate the betrothal day, June 24. Days later, Marie took Irène to Brazil.

The University of Rio de Janeiro had invited Mme. Curie to give a series of lectures on radioactivity. Marie, who routinely accommodated foreign radioactivists in her lab, recognized a parallel need to deliver her knowledge directly to scientists abroad. She chose Irène to travel with her not only as a traveling companion, and not just to test the fiancée bond Irène had forged with Frédéric, but as an essential assistant and demonstrator in at least a dozen formal presentations. This might be the last trip they would ever take together, just the two of them. They sailed from Marseilles on June 30.

"The cabin is airy and comfortable," Marie wrote from the *Pincio* to the daughter left at home, "but there's no hot water in the bath. We may be able to carry some down in jugs, although this possibility is limited by the steepness of the final staircase. Apart from that, we are well situated and the boat is very steady." They were coasting along eastern Spain, and she planned to post her letter at the stop in Valencia. "The food is good, rather too much of it, though, which enables us to skip the buffet items most likely to spoil in warm weather, such as meats and eggs." After leaving the last Spanish port, they encountered no other vessels during the two weeks their ship plied west and south across the empty, open ocean. Their shadows shrank to nothing under the high sun of the tropics. The nights filled with stars arranged in novel constellations.

"We are in sight of the coast of Brazil and should reach Rio around noon," she told Ève on July 15. Even before she and Irène could disembark at Guanabara Bay, a welcoming party motored out to greet them in a launch bedecked with flowers. The dignitaries aboard included the French ambassador, the rector and several professors from the university, and members of a committee representing the women of Brazil.

Once landed with Irène in their well-appointed hotel room, Marie found a nearby beach where they could swim and sunbathe early in the morning, before reporting to the university lecture hall. Their Brazilian hostesses accompanied them to official entertainments, and happily led them on recreational walks and drives in the hills above the noisy city, where the visitors marveled at orange and banana trees and great green expanses of utterly unfamiliar plants. Some of their congenial new friends were scientists in their own right, others active in government, and all supported the country's nascent suffragist movement, the Brazilian Federation for Female Progress.

Irène wrote to Frédéric every day. He was not as frequent or voluble a correspondent as she, but clear in communicating how much he missed her. "I go to the laboratory, and it is a sham not to find you there," he said. "I shall not return to it with any happiness until you are there again."

After Rio, Marie and Irène gave further presentations in São Paulo and at the University of Minas Gerais in Belo Horizonte, the site of a medical school with a radium institute. They received several pressing invitations to lecture and teach in Argentina as well, but Marie refused these. She did not want to extend the already lengthy South American tour, partly because the shorter days and colder temperatures in Buenos Aires, where winter now prevailed, would deny her the dose of sunshine and warmth she required for her health. Moreover, it was high time to reunite the young lovers.

As soon as Irène returned in September, she initiated Frédéric in the pleasures of l'Arcouest. The rented cottage that defined summertime for her now belonged to the Curies. Marie purchased it in 1924 with part of her government pension, and then acquired a second property—a plot of vacant land closer to the beach and the dock—in 1925. If Frédéric fit in with the l'Arcouest community as comfortably as Irène hoped, then she and he might build their own place someday. Luckily, Frédéric's skill as a fisherman endeared him to the Brittany locals and summer folk alike.

The wedding of Irène and Frédéric, like the wedding of Marie and Pierre, was a small civil service. It took place on Saturday morning, October 9, in the municipal offices of Paris's fourth *arrondissement*, attended only by immediate family and closest friends. A luncheon chez Marie followed. That afternoon, instead of bicycling off on a honeymoon, the newlyweds resumed their experiments in the lab. The next day, when Marie departed for Denmark on another round of lectures, Frédéric moved into Irène's room in the all-female lair on the quai de Béthune.

The start of the new academic year brought Irène the first foreign scientist who had come expressly to study under her. In the past Irène had tutored any number of visitors and newcomers, notably Fred, but none of them had sought her out. Hungarian-born Erzsébet Róna, an experienced radioactivist with the Vienna Radium Institute, had requested specific instruction in Irène's techniques for separating, purifying, and concentrating polonium. Mlle. Róna shadowed her younger mentor for two months to gain mastery, and left Paris with a modicum of polonium as a parting gift from the Curies.

In December Irène heard from her previous mentee and collaborator Nobuo Yamada. He had been back in Japan just two weeks, his letter said, when he suddenly fainted. Since then a strange malaise had confined him to bed. "The cause of the illness still isn't clear," he wrote, though he assumed it had to do with radioactivity exposure. "It is certain that I was very tired after the long stay abroad, but also there was a poisoning from the emanations." This was his own diagnosis, based on his symptoms of chronic stomach distress and crippling fatigue.

In Tokyo, where Yamada lay ill, doctors still had little or no experience with radioelements, either as boons or as threats to health, and not a single article about radium poisoning could be found in the Japanese medical literature. Therefore, Yamada hoped Irène could provide such information from other sources, allowing him to compare the progress of his disease with those reports.

Expressing her concern, Irène did her best to comply.

Chapter Twenty-Four

ÉLIANE

(Polonium *bis*)

———

IN THE MARRIAGE of Irène and Frédéric, Irène was the older, better-known, and more scientifically savvy partner. Although legally now "Mme. Joliot," she could hardly be expected to change her name, given her professional position and the number of papers she had already published under the byline of Mademoiselle Irène Curie—not to mention the aura of the Curie name in the field of radioactivity. Frédéric understood these realities as well as she did. Thus the couple's first joint report to the *Académie des Sciences*, presented by physicist and family friend Jean Perrin, was coauthored by *Madame* Irène Curie and Monsieur Frédéric Joliot.

The early months of their married life were marred by illness. Frédéric suffered an attack of appendicitis in December 1926 and underwent surgery in January. In February, after Irène announced her pregnancy, her doctors urged her to rest—not merely because she was expecting, but because her chest X-rays showed signs of tuberculosis. Since Irène's prescription for rest dovetailed with Frédéric's, they quit Paris in March for Porquerolles off the Côte d'Azur.

The island was so small that one could traverse it or circle it with ease, Irène wrote Marie, but she and Fred felt so tired and lazy that they rarely went walking at all. They had not even tried to work. Instead, they "ate like horses" and gathered branches from the pine woods or driftwood on the shore to stoke the woodstove in their cozy room at the *pension Sainte-Anne*.

Marie replied with news of the lab, where an experiment they had left in her care was producing "devilish" results. Irène suggested that Marie "choose a good moment, in the middle of all this polonium," to take a break and join them on Porquerolles. Frédéric, more troubled by Marie's apparent distress at the results than by the experiment's failure, seconded the invitation. "Permit me to intervene to insist that you join us here," he wrote. "You must not (excuse my tone!) let yourself fall ill by refusing to rest."

Marie found the *pension* and the island quite to her liking. Better yet, she saw that Irène's lips and cheeks had turned rosy, and she showed a mother-to-be weight gain of about five kilos. "We are installed for a few days in Menton," Marie informed Ève from one of the coastal towns they explored, "as I wrote you yesterday. I sent you a telegram, too. You must also have received the box of candied fruits shipped from Hyères. So you know I don't forget you."

Ève was establishing herself as a music critic, writing for several periodicals. She had bought a car and rented a studio where she could comfortably entertain friends from her world of artists, but she continued living with her mother. The situation suited them both. Ève thought Marie would be too lonely living alone, now that Irène and Frédéric had moved to their own apartment in the rue Froidevaux, and Marie thought Ève, at twenty-two, was too young to live alone.

In Paris, Irène and Frédéric came to dinner with Marie and Ève three or four times a week. On these occasions, science again dominated the conversation at table, except that now Marie and Frédéric did most of the talking, so that Irène complained she had to struggle to get a word in.

The newest face at the Curie lab belonged to Éliane Montel, a 1923 graduate of the *École Normale Supérieure de jeunes filles de Sèvres*. She had taken classes there with Paul Langevin, who recognized her talent and recommended her to Marie.

Madame's first Sèvrienne, Eugénie Feytis Cotton, had at last attained her doctoral degree in physics in 1925, the same year as Irène, her former babysitting charge. By the time Eugénie, as a mother of three and head of physics instruction at Sèvres, found the time to summarize results of magnetism studies begun more than a decade earlier at the Sorbonne and in Zurich, her experiments had long since been superseded by other scientists using improved techniques. Nevertheless, her Zurich mentor, Pierre Weiss, read her thesis and suggested ways to bring it up to date. She successfully defended the revised dissertation at the University of Strasbourg, where Weiss had moved following the death of his wife. In the course of these events, Eugénie introduced the widower to her own former student, Sèvrienne Marthe Klein, one of Mme. Curie's wartime instructors of X-ray technicians. After Marthe became the second Mme. Pierre Weiss and stepmother to his six-year-old daughter, she continued teaching physics to girls.

In the Sèvres tradition of educating women for careers in science, Éliane Montel taught at a female academy while assessing the penetration of polonium in lead at the *Laboratoire Curie*.

———

"I DON'T KNOW how the time passes," Irène wrote to her mother in the summer of 1927. "If I didn't have my journals and the daily arrival of the newspaper to recall me to reality, I would lose any notion of what day it is." She felt too tired and too heavily pregnant to man the oars in a rowboat on the Brittany coast, so she and Frédéric ceded the house in l'Arcouest to Marie for the time being. The expectant couple ventured only as far as a cottage in Brunoy, just outside Paris, for their vacation.

Irène slept "like an animal in hibernation." Frédéric fished in the river Yerres and regularly took the train to Paris to check on progress at the lab. On August 2 he found one of the staff members looking unwell. Sonia Slobodkine—Sonia Cotelle, now that she had married—was usually the most cheerful person in the whole Radium Institute, but she had endured days of stomach trouble, she told him, and had shed alarming quantities of her hair. Frédéric pressed her to leave immediately for an indefinite period of fresh-air therapy.

On hearing Fred's account, Irène thought Sonia might have swallowed some polonium while pipetting a solution. On second thought, while recounting the incident to Marie, she deemed it equally likely that Sonia's "present condition has no connection with polonium." Still, Irène added, "she feels very uneasy, which is understandable."

This was the state of cognitive dissonance in which all radio-activists lived. On the one hand, they appreciated the destructive power of the materials they handled daily. On the other hand, any worrisome symptoms they developed could be alleviated by something as simple as a brief holiday. Always the allure of the radioelements, the camaraderie of the lab, the chance of making a new discovery drew them back to work.

Irène could blame her own current fatigue on the predictable weariness of pregnancy. She gave birth on September 19, a week after her thirtieth birthday. She was the same age Marie had been when Irène herself was born. Like Marie, Irène reserved a notebook for observations about her daughter, in which she described the newborn Hélène Joliot as "a small baby, not too red, with a little blond hair," and somewhat resembling "a young carp." Unlike Marie, who returned to her magnetized steels within four days of her firstborn's arrival, Irène required a lying-in period that lasted a fortnight. She left the maternity clinic well aware that she had not yet regained all her strength.

———

"I ARRIVED IN BRUSSELS without mishap," Marie wrote Ève on October 25, at the start of the fifth Solvay Council, "a little tired, naturally, and found here the most brilliant company from the scientific point of view." The group of thirty included Albert Einstein, Max Planck, and Niels Bohr. "I take such great pleasure in speaking of new things with all these lovers of physics."

At the first Solvay gathering, sixteen years before, Planck's quantum—the packaging of energy into discrete bundles—had been a novelty, at odds with every tenet of classical physics. By 1927 the quantum had become a pillar of science, and had cleaved the universe into two seemingly irreconcilable halves. In the familiar realm, molecules and organisms were held together by chemical bonds and electromagnetic force, and obeyed the same law of gravity that controlled the motions of planets and stars. But different rules and forces prevailed in the subatomic domain, where waves could behave like particles, and particles behaved like waves. New math had been necessary to explicate the atom, and the mathematicians who met that need—Louis de Broglie, Erwin Schrödinger, Max Born, Werner Heisenberg, Paul Dirac, Wolfgang Pauli—were all present in Brussels for these discussions.

Marie, still the council's only female physics insider, had kept up with recent developments. She presented remarks she had prepared in advance regarding experiments by Arthur Holly Compton of the University of Chicago, one of the youthful new attendees. Just as the alpha and beta "rays" of radioactivity had turned out to be tangible particles, namely helium nuclei and electrons, light rays, too, according to Compton, could be resolved into discrete particles. He had demonstrated that when a quantum of light energy, or "photon," struck an electron, the two rebounded like colliding billiard balls.

Conversations devoted to "Electrons and Photons" quickly escalated to debate. Hendrik Lorentz, who had presided over every council since the first one in 1911, strove to maintain order, but often several participants called out at once in their excitement, and in their own languages. Arguments between Einstein and Bohr over the interpretation of quantum theory spilled out of the conference room into the city streets, often carrying on through the dinner hour and late into the night.

Soon after Marie got home from this rarefied fray, she learned that Irène's Japanese protégé, Nobuo Yamada, had died in Tokyo on the first of November. Tokyo Imperial University, the letter said, had promoted the thirty-one-year-old Yamada to the rank of full professor while he lay unconscious in the final weeks of his life. Doctors attributed his death to a brain tumor.

Although Yamada had encountered every type of radiation in the Curie lab, his two years of exposure could not be conclusively linked to his illness or death. A preexisting or otherwise unrelated problem might have been to blame. Irène's own ongoing fatigue and anemia, however, were becoming harder to explain away.

"Irène doesn't feel well yet," Marie wrote her brother, Józef, in early December. "She still doesn't have enough erythrocytes [red blood cells]. She will be leaving soon for two weeks of winter sports, and hopes that this stay in the mountains will be good for her anemia."

In January, before Irène set off with Frédéric for the Alps, news broke of a new court case concerning luminous-dial painters in the United States. Five sick and dying young women claimed that the practice of putting radium-laced paintbrushes in their mouths had ravaged their bodies. They were suing the U.S. Radium Corporation for help with the extraordinary medical expenses they incurred on account of their unique disabilities. In addition to the "radium jaw" that allegedly caused their teeth to fall out, they sustained other bone destruction in their feet, knees, hips, and back, treatable only by orthopedic braces and surgery.

Reporters contacted the discoverer of radium for comment. Mme. Curie believed the women deserved compassion, justice, and compensation. "I would be only too happy to give any aid that I could," she said. Regrettably, however, "there is absolutely no means of destroying the substance once it enters the human body." She meant there was no way to remove it from the victims' bodies. Radium could not be destroyed by any means other than its own self-destruction, at a pace so slow as to far outlast any human lifetime.

Chapter Twenty-Five

ANGÈLE (Bismuth)

———

TO IRÈNE, the Curie lab at the *Institut du Radium* had long felt as much like her home as her place of work, with her mother presiding over both. Since her marriage to Frédéric, the lab had become an extension of the marital bed as well. In the autumn of 1928, Irène's circle of lab intimates widened to include her best friend, Angèle Pompéï, who came to work alongside her.

The two young women had met in their early twenties on a geological field trip to volcanic regions of the Auvergne. Irène signed up seeking open-air exercise and science; Angèle was satisfying a requirement for her physics teaching degree from the *École Normale Supérieure de jeunes filles de Sèvres*. Together they mounted further adventures of their own design, from summer hikes in Angèle's native Corsica to winter treks in the French Alps. After Angèle moved to Algiers in 1923 to teach in a girls' school there, Marie and Irène visited her, stayed in the quarters she shared with two other teachers, and toured the surrounding countryside by rail and camel.

As soon as Angèle completed her service in Algeria, she returned to Paris to teach physics and chemistry at the *Lycée de Saint-Germain-en-Laye*, and entreated Mme. Curie for a part-time research post at the Radium Institute. Marie, pleased to accept another eager Sèvrienne into her fold, assigned Angèle to a team studying beta emission from radium E.

Angèle suspected it might take her a while to learn the alphabet soup of names marking successive stages of transmutation, though their utility was easy to see. "Radium E" surely made convenient shorthand for "radioactive isotope of the element bismuth, with a half-life of five days." And "radium E" further served to distinguish this radioactive isotope of bismuth from the other, heavier one, "radium C," which formed earlier in the decay chain and had a half-life of less than half an hour. As yet, no one could explain the great differences in stability between isotopes of the same element.

Because Angèle earned a teaching salary from the *lycée*, and lived frugally with Madeleine and Marie Elichabe (the same pair of sisters who had been her housemates in Algiers), she did not request a Curie-Carnegie scholarship or other financial support. All she wanted was opportunity and a place for experimentation, both of which abounded at the *Pavillon Curie*. The building, spacious to begin with, had lately been expanded by the addition of a new wing.

In the cramped confines of the old Curie lab in the Annex, Marie had nearly turned away Ellen Gleditsch for want of space. Ellen, who still lacked a proper laboratory of her own, was awaiting completion of the new chemistry building at the Royal Frederick University. After careful study as a member of the planning committee, she had petitioned for ten rooms devoted to radiochemistry, including laboratories, a darkroom, a weighing room, and an office for herself. As she looked forward to those accommodations, the retirement of her colleague Heinrich Goldschmidt raised the possibility of her advancement to the rank of full professor. Ellen had been an associate professor for thirteen years. She could boast far more laboratory experience than any of the other candidates for the coveted position, as well as an unparalleled reputation as a lecturer. Working against her were the facts that she was fifty and female.

In March 1929, shortly before the faculty board voted on the chemistry professorship, Ellen left Norway to tour the United States in her capacity as president of the International Federation of University Women. The IFUW had grown to thirty-seven thousand members in thirty-one countries. It currently empowered twenty-two young women to do as Ellen had done—to travel abroad and profit from contact with foreign mentors. By lecturing for six weeks in venues from New York to San Francisco and Minneapolis to New Orleans, she hoped to raise additional fellowship funds.

Portrayed in the American press during her 1913 visit as a "pretty little woman" with "sweet and smiling lips," Ellen was described this time by the *Oakland Tribune* as "a world famous scientist who can look Einstein's formula in the eye, without blinking." In technical talks she gave at colleges including Stanford and Yale, she detailed her ongoing research regarding the isotopes of chlorine. Although chlorine was not a radioelement, it existed in two distinct forms, one with atomic weight 35 and the other with atomic weight 37. Ellen was testing a variety of chlorine-containing compounds, such as sea water from different oceans, to see whether the two isotopes of chlorine always occurred in the same proportion, and so far it seemed to her that they did.

When speaking to nonscientific audiences, she stressed the rewards of international exchange. "A student returns from such a stay abroad greatly enriched," she maintained, "not exactly in gold, but in noble goods."

Interviewed about her second impressions of America, Ellen said she thought barriers to women's lab access had fallen in the fifteen-year interim. "All of the younger scientific men have ceased to be either resentful or condescending towards their female colleagues," the *Tribune* quoted her as saying, "and the older and more conservative ones are dying out, or are submitting to the new conditions."

Ellen was thousands of miles from Oslo when the faculty board voted ten to three in her favor, granting her the much-desired promotion. But then a dean who preferred one of the junior candidates questioned certain steps in the selection procedure and succeeded in overturning Ellen's election.

Stopping at Paris on her way home in May, Ellen asked the director of the *Laboratoire Curie* for a letter of recommendation. Madame's word, coupled with the strong support of Professor Kristine Bonnevie and Ellen's original champion, Eyvind Bødtker, restored her victory. "It's been done," Ellen wrote Marie on June 26. "Three days ago I was appointed professor." At a celebration of her achievement arranged by the Norwegian organization of university women, Ellen told her well-wishers:

> To me it is absolutely indifferent whether a piece of work is carried out by a small woman in Bulgaria or by a tall man in America, as long as it is done well. And this is what we have to do: do things so well that no one would dare to say, "this is good work for a woman," but that everyone will say, "this is good work."

MARIE ALSO REVISITED the United States in 1929, begging a second indulgence of the American people. She needed another nugget of radium for the new research institute and hospital nearing completion in Warsaw. When she queried Missy Meloney about the chances of getting one more gram as a gift, Missy rang up several members of the 1921 Marie Curie Radium Fund Executive Committee. One of these women, geologist and philanthropist Lou Henry Hoover, was now the country's First Lady, following the election of her husband, Herbert, to the presidency.

Marie voyaged alone aboard the "floating palace" of the *Île-de-France*. With no protective companions, and knowing herself to be an object of curiosity, she took the air on the uppermost deck. There, she wrote Ève, the force of the wind sometimes stopped her in her tracks, but she chose the exposed, windy walk over the more crowded covered promenade. By staying out of the salons, often keeping to her cabin, she avoided awkward encounters with fellow passengers. These included entertainer Maurice Chevalier, whose performance during the crossing met with her approval.

The moment the ship arrived in New York, on October 15, Missy was at Marie's side. She had scheduled several social gatherings in or near her apartment in the city, and once snuck Marie downstairs by the service elevator to escape reporters thronging the lobby. When it came time to move on to the next hostess, in Manhasset on Long Island, Missy organized a car ride with a police escort. They traveled by train to Dearborn, Michigan, for the fiftieth anniversary commemoration of Thomas Edison's lightbulb, and to Canton, New York, where Marie dedicated a new science building at St. Lawrence University. The main event—the brief but impressive radium presentation ceremony—took place at the National Academy of Sciences in Washington, DC, a day after the disastrous decline, on October 29, 1929, of the New York Stock Exchange.

In lieu of an elaborate treasure chest or a symbolic gold key, President Hoover handed Mme. Curie a bank check for $50,000. This sum represented the going price for a gram of radium. Increased supply, primarily from the Haut-Katanga region of the Congo, had cut the 1921 cost in half. A ton of African pitchblende yielded thirty to forty times more radium than a ton of Colorado carnotite—the source of Mme. Curie's previous gram of radium.

At dinner that night in the White House, Marie grew positively chatty with President Hoover, and accepted two tiny elephant statuettes—the totem of his political party—to take home as souvenirs. She also collected a trove of scientific instruments on her travels, to replace antiquated ones in the Curie lab. A Leeds Northrup galvanometer, considered the *dernier cri* in electrometers, received special mention in a note to Irène.

"The financial catastrophe that erupted here seems calmed now," Marie wrote home to Ève on October 31, when all but one of her official duties had been discharged. The quiet pause allowed her to reflect on the genuineness of her friendships in America. "I regret that the distance is

Mme. Curie with President Herbert Hoover

too great to allow easy exchange," her letter continued, not just from Europe to the States but even from one state to another.

Missy, who likewise regretted the great gulf between them, sent Marie an address book as a Christmas gift, with the initials MSC stamped in gold on its green leather cover. Inside, Marie found the names and addresses of all her American friends already penned in.

Chapter Twenty-Six

ISABELLE and ANTONIA
(Thallium)

———

"THE OLDER ONE GETS," Marie observed in a reflective mood, "the more one realizes that knowing how to enjoy the present moment is a precious gift, comparable to a state of grace." It was better to enjoy today *today*, she told her children, than to look back and savor it later, or to put off enjoyment till some distant tomorrow.

At sixty-two, having lived so much longer than either her mother or her husband, and with her future foreshortened accordingly, Marie spent her spare time preparing a revised second edition of her two-volume treatise on radioactivity and overseeing completion of the new radioelements factory at Arcueil, six miles from central Paris. The text encapsulated the history of the field she had named and still toiled in; the factory ensured its future.

Radioactivity itself was exquisitely time sensitive, with each radio-element defined by its rate of diminishment. Had she known, in the early days of her discoveries, the difference in longevity between polonium and radium, would she have saved her native country's name for the more durable element?

The clocklike regularity of radioactive decay was what enabled scientists, despite their puny human lifespans, to extrapolate the sixteen-hundred-year half-life of radium and the multi-billion-year half-life of uranium. Even the lifetimes of individual atoms could be calculated: The average radium atom, though it had the potential to explode at any moment, endured twenty-five hundred years before expelling an alpha particle and turning to radon. In contrast, no single atom of polonium ever survived for more than a few months' time.

The radium standard that Marie had fabricated in the summer of 1911—a petite glass tube containing twenty-two milligrams of her purest radium chloride—still served as the world's arbiter of radioactivity. It resided, as per agreement, at the *Bureau International des Poids et Mesures* in Sèvres, but traveled to the *Laboratoire Curie* whenever required to test a secondary standard bound for some new member of the global radioactivity community. By 1930, half a dozen secondaries had been prepared in Vienna, tested in Paris, and exported with certificates signed by Marie Curie, Ernest Rutherford, and Stefan Meyer, officers of the International Commission for the Radium Standard.

Marie personally tested the first few secondaries, by comparing their degree of gamma radiation to that emitted by the international standard. Later she trained Renée Galabert to conduct the tests. Renée, who joined the lab in 1919, had advanced to head of the Measurement Service in 1921, the year Madeleine Molinier left to have a baby.

Thanks to the long half-life of radium, the international standard remained virtually unaltered after two decades of use. Each year saw it shrink by only one-thousandth of its initial mass. Still, time wrought important changes. The moment Marie sealed the glass tube, a predictable number of radium atoms transformed into radon, and the radon gas changed in its turn, yielding a series of solid daughter products. At the end of one month, a state of equilibrium prevailed inside the tube: the amount of new radon forming exactly balanced the amount dissipating.

The condition of radioactive equilibrium figured crucially in defining the "curie," the unit of measure named in 1910 in honor of Pierre: one curie signified the quantity of radon (then called "radium emanation") in equilibrium with one gram of radium. At the time, other radioactivists wanted to base the measure on a much smaller weight of radium—a milligram instead of a gram. Marie had insisted on a whole gram, however, and won her way. Perhaps she had been showing off, as she alone actually owned a full gram of radium. More likely she meant to set Pierre's namesake unit on a generous foundation. As a result, measurements had been expressed ever since in *millicuries*.

In 1930 the international commission broadened the definition of the curie so that it was no longer tied exclusively to radon. From now on, one curie would equal the amount of any decay product in the uranium-radium family—polonium, say, or thallium—in equilibrium with a gram of radium.

Over time the comings and goings of personnel in the Curie lab had also achieved an equilibrium. The outflow of radioactivity-wise students, assistants, and independent researchers invited an influx of others equally eager to learn. In February 1930, Isabelle Archinard arrived from Geneva, on the advice of her thesis advisor, Charles Eugène Guye, who held the University of Geneva physics chair once offered to Pierre. Isabelle had also been pointed toward the Curie lab by Jean d'Espine, a veteran of five years' collaboration in Paris with

Prof. Ellen Gleditsch, 1935

Marie, beginning in 1921. The Swiss d'Espine would have stayed longer at the Radium Institute, but he contracted tuberculosis and went home in 1926 to take the cure.

Some lab alumnae—Ellen Gleditsch and Alicja Dorabialska in particular—returned periodically to undertake new studies in radioactivity or to spell Madame when she was teaching or lecturing abroad. Ellen, now Professor Gleditsch, had secured only half the space she requested in the long-awaited new chemistry building at the Royal Frederick University. Instead of ten rooms *en suite*, just two on the second floor and two in the basement were assigned to radiochemistry. No sooner had the chemists occupied the building than a plan emerged to move both the chemistry and physics departments to a shared new space at the university's Blindern campus on the outskirts of Oslo. In the spring of 1930, Ellen again consented to serve on the planning committee.

"I have dangerously many interests beside my science," she confessed to the student body at the 1930 jubilee. "I enjoy fiction, theater, and music, although not cinema. I am a member of the Female Students' Choir and on its board. In recent years I have participated a little in international work, the peace movement, and intellectual cooperation." Indeed, as part of her contract with the IFUW, Ellen had interacted with committees and subcommittees formed by the League of Nations. "I love sports and outdoor life, am a member of the Female Students' Skiing Club, and know no better kind of summer holiday than a sojourn in the Norwegian mountains. I have lived long enough in France to appreciate that cooking is a great and important art, and I am known in many countries for my omelets."

———

WHILE WRITING HIS doctoral dissertation on "The Electrochemical Study of Radioelements," Frédéric Joliot earned extra money to support his growing family by teaching part-time at the *École Pratique d'Électricité*. On receipt of his graduate degree in mid-March 1930, he and Irène conceived new experiments together and also built and tinkered with the equipment needed to execute their ideas. Although the Radium Institute employed two full-time mechanics, an electrician, and a glassblower, Frédéric relished doing the mechanical work himself. He excelled at glassblowing, too. As he and Irène had predicted when they married, their contrasting natures and modes of thinking complemented each other well. Their hired nanny, Mme. Blondin, looked after Hélène during their long days at the lab.

Irène was still not well. In June she availed herself of the mountain air at Notre-Dame de Bellecombe, a ski resort near Mont Blanc in the Savoie region. A concerned Marie went with her. Organizing themselves around the daily rain showers, they walked for hours every afternoon. They ate, slept, took hot baths, and largely ignored the writing and editing work they had brought along. Ève sent Marie a book to read for relaxation. It was Colette's latest, called *Sido*, a memoir about her mother. Ève deemed it "a masterpiece, every sentence of which is to be savored," and Marie could only agree. Meanwhile Frédéric commuted regularly between Paris and Brunoy to spend time with Hélène. "The visits from her father," Marie wrote Ève, "must comfort the dear child while her mother is so far away."

Irène had built up a "depth of fatigue" that would not ebb easily, Marie feared, due to the persistent anemia. Nevertheless, the current regimen seemed to be doing her some good.

A plea from Paris abruptly ended Marie's part of the mountain respite. Her colleagues Georges Urbain, Jean Perrin, and Émile Borel begged her to join them in a July 8 meeting with Gaston Doumergue, the president of France, to discuss the newly established funding agency for scientific research, the *Caisse Nationale des Sciences*, and she could not refuse. One of the first Caisse grants was awarded to the promising young Frédéric Joliot. The state-sponsored windfall would allow him to quit his sideline teaching job and devote himself totally to research.

With Irène still away, Frédéric took Hélène to l'Arcouest on vacation. "I wrote Irène to recommend she get a blood count as soon as she returns," he informed Marie. "If you are in Paris then, please insist that she listen to me." In other news, he said Hélène had formed a marvelous friendship with Charles Seignobos, the elderly Sorbonne historian hailed by the summer regulars in l'Arcouest as "*le capitaine*" of all sailing and social activities. Now Seignobos let the littlest member of the company gather flowers in his garden.

August took Irène back to her family and back to Brittany. "I have done a little bit of work, very little," she wrote Marie from l'Arcouest, "but if we have bad weather, I think I'll get better motivated." She was planning the students' projects for the coming academic year, matching their interests with their skills and experience.

The start of the fall term was to bring Dutch chemist Antonia Korvezee to the Radium Institute from the Technical University in Delft, where she had recently completed her doctoral degree and won a scholarship to further her education abroad. Although Irène did not keep a running count, Mlle. Korvezee would be the thirty-fifth woman to enter the Curie lab since Harriet Brooks found her way to the old Annex in 1906.

Marie, in Cavalaire, sent bonbons to Irène for her birthday, followed by a teddy bear and a construction set for Hélène's birthday.

"I had a toy chest built for Hélène," Irène reported while still in l'Arcouest in mid-September, "but I had it made a little too large. And the result is that Hélène, after putting her toys inside it, climbs in to play with them. Making off with the chest just then would capture the baby and all her belongings in one fell swoop."

That same month saw the death of Kazimir Dluski, Marie's protective "little brother-in-law" who had looked out for her during her student days at the Sorbonne, and who had shared Bronya's life for forty years.

"My siblings are in reasonable health," Marie wrote Ève from Warsaw, not long after the funeral. They lived "in a circle that tightens as they advance in age, and in which the loss of Uncle Dluski has left a sorrowful gap." Bronya was coping with her grief by concentrating on her professional duties, readying the new radium institute and hospital. "What I fear most for her," Marie said, "is the moment when she will need to curtail her activities, because then the past will come round to haunt her."

Chapter Twenty-Seven

BRANCA (Boron)

A STEADY RAIN beat on the train carrying Marie and Ève south to Spain in April 1931. They saw little of the passing countryside but found entertainment in conversation with physicist Blas Cabrera of the University of Madrid—and with his companion, who spun tales of traveling the world by zeppelin—till they crossed the Pyrenees and the skies cleared.

"When we arrived at the station in Madrid," Marie wrote to Irène, "we were welcomed kindly with a magnificent bouquet of red carnations, then we collapsed at our lodgings in the women's residence, only to be frozen solid because the heat wasn't working." The present moment, she said, found her "next to a warm radiator, without shivering," and hoping to avoid a second night of poor sleep troubled by nightmares. "I dreamt I had fled Madrid in secret, and, realizing what I had done and the inconvenience it would cause, I was trying desperately to return."

Ève added her own colorful account of the frigid night in a post-script: "I screamed, broke furniture, and committed acts of war until someone finally turned on the heat."

Abundant sunshine on the following day, luncheon with Mme. Cabrera, and a tour of the physics laboratories did much to warm the travelers. Naturally Ève was not expected to assist in any scientific demonstrations, as Irène would have done, but simply to escort their mother through a series of guest lectures and social engagements in Madrid, Toledo, Granada, Málaga, Almería, Murcia, Alicante,

Valencia, and Barcelona. King Alfonso XIII, whom Marie met in 1919 while teaching her radioactivity course to his subjects, had just been deposed, and the former monarchy was changing radically to a republic.

"We are very well received," Marie wrote Irène again on April 24, after giving two presentations in Madrid. "We meet people rejoicing in their young republic, and it is very moving to see such confidence in the future among the youth, and among many of their elders as well." The gracious Ève had made "conquests, as usual" during student encounters and visits to the French and Polish embassies.

Once they left the capital, their busy itinerary, coupled with the slow pace of mail, kept them a step or two ahead of letters from home. "I am without news of you and of the Laboratory," Marie rued to Irène, "and this makes me feel uneasy, as though I were on another planet."

Of course she knew she could trust Irène, André Debierne, Sonia Cotelle, Catherine Chamié, and Léonie Razet to run things in her absence. But it had become more difficult for Marie to be absent from the lab, even as the reasons for her absences multiplied.

"I arrived here this morning," she notified Ève from Geneva in mid-July, at a meeting of the International Committee on Intellectual Cooperation, "after a pleasant enough journey with an adequate amount of sleep. I dreamt, I don't know why, that the public had invaded the wagon-lit, making it impossible for me to dress myself in the morning." Her committee duty was getting to be "a very heavy task," she allowed, but one that she deemed indispensable and worthy of personal sacrifice.

She got back to Paris just in time to serve as honorary president of the Third International Congress on Radiology, chaired by Dr. Antoine Béclère, and held at the Sorbonne the last week of July. In the years since the Great War, X-rays and radioactivity had become integral to medical diagnosis and treatment. Physicians were now looking beyond ampoules and needles of radon gas implanted inside the body to "teletherapy," or the irradiation of cancerous tumors from without, using radium positioned somewhere near the patient.

A contingent of doctors from the United States took the occasion of the Paris congress to bestow on Mme. Curie the gold medal of the American College of Radiology.

No Solvay Council was scheduled for 1931, but an International Congress on Nuclear Physics drew Marie and most of her Solvay associates to Rome in October. Enrico Fermi, a new name in the nuclear community, organized the congress to signal Italy's serious entry into this burgeoning field of research. "I don't know everyone here," Marie told Irène on the thirteenth, though at least twenty of the fifty participants were acquaintances, among them Niels Bohr, Max Planck, Arthur Holly Compton, and Werner Heisenberg. In addition to Mme. Curie, usually the only woman at such get-togethers, Fermi had invited Lise Meitner of the Kaiser Wilhelm Institute for Chemistry in Berlin. The Vienna-born Fraulein Meitner, a co-discoverer of the element proto-actinium, was the first woman to be elected physics professor at the University of Berlin—the first, in fact, in all of Germany.

"I try to follow the reports as much as possible," Marie continued in her letter to Irène, "which is not always easy, considering the extreme technicality and above all the lack of clear elocution among certain speakers. I think I'll have a few words to say when the discussion turns to radioactive phenomena." She had not taken in the sights of Rome, nor did she expect to. "I have little else to tell you at this point, except that Bohr strongly insists on the impossibility of actually applying quantum mechanics to the interior of the nucleus. Love, Mé."

Niels Bohr had successfully connected quantum mechanics—the new understanding of events at the atomic and subatomic scale—to the exterior of the atom, that is, to the orbiting electrons. As he envisioned the atom, electrons could occupy only certain orbits, representing discrete energy levels. When an electron absorbed a specific quantum of energy, it instantly shifted to a higher orbital shell, and when it dropped down again to its accustomed level, it emitted that quantum of energy as a specific wavelength of light. This activity accounted for the colors of lines in the visible spectra of elements.

Marie, a close witness to the growth of quantum mechanics, well appreciated the theory's departures from classical physics. As a human being in the realm of ordinary experience, she had traveled from Paris to this meeting in Rome along a continuous stretch of railroad tracks. At the scale of an electron in the subatomic realm, where discreteness trumped continuity, she would have been either in Paris or in Rome, and never anywhere in between.

Each orbital shell of Bohr's atom had its own population limit: two electrons at most could inhabit the lowest one, eight apiece in the next two, and eighteen in each of the following two. Atoms with outermost shells filled to capacity could not engage in chemistry with other atoms. This limitation explained the inactivity of the inert, or "noble," gases. The completion of each electron shell coincided with the end of a period, or horizontal line of elements on the periodic table.

The nucleus within the electron shells—the interior of the atom, the seat of radioactivity—remained elusive. Known only as the locus of positive charge, it somehow managed to expel negative electrons in the course of beta decay. Some of the scientists present in Rome had already tried to poke at the nucleus by various means in the hope of exposing its secrets. Perhaps the most riveting talk Marie heard was the one presented by Walther Bothe, director of the Institute of Physics at the University of Giessen. He and a student collaborator, Herbert Becker, had used beams of alpha particles from polonium to bombard targets made of lightweight elements such as lithium, beryllium, and boron. Bursts of penetrating radiation fled the struck targets with enough energy to pierce a piece of lead several centimeters thick. The emitted radiation resembled gamma radiation, in that it carried no charge, and its measured energy exceeded that of the bombarding alpha particles. Apparently, the target materials, although not radioactive, had undergone some type of nuclear decay. Neither Bothe nor anyone else could explain this strange outcome.

MARIE RETURNED FROM her international travels to the international milieu of the *Institut du Radium*. The forty men and fifteen women now working in the expanded *Laboratoire Curie* represented nearly every European country, as well as Russia, China, and India. "*La patronne*" had begun holding weekly meetings to afford researchers a chance to report their progress or bemoan a setback. Her newest trainee, Branca Edmée Marques, came from the University of Lisbon, having taught courses for six years there in physical, organic, and analytical chemistry. At age thirty-two, she was one of several talented young career scientists selected to staff the new Portuguese radiological institute, and had been awarded a government fellowship to acquire the necessary experience abroad. Branca Edmée Marques chose to learn the measurement of radioactivity and the chemistry of radioelements in Mme. Curie's lab, even though her husband and former professor, António Sousa Torres, could not accompany her to Paris. She brought her mother instead.

Irène had recently shown several years' worth of her chest X-rays to a new doctor, who reassured her as to the condition of her lungs. It seemed her tuberculosis was in remission, though anemia continued to weigh on her. Thus far that year she had sought relief at three mountain retreats, beginning in April at the Jura range on the Swiss border. In August, upon her arrival in the Chartreuse Mountains just north of Grenoble, she complained of feeling "very tired and incapable of walking for as much as an hour." She rallied as the days passed, however, and moved on to meet her colleague and hiking pal Angèle Pompéï at le Monêtier-les-Bains, a resort area that offered them mineral baths as well as the beneficial altitude of the Hautes-Alpes.

Irène rested some more in September with her husband and daughter at l'Arcouest. She had been warned, while pregnant with Hélène, that she should not attempt to have a second child, given the precarious state of her health. Nevertheless, she was expecting again when she and Frédéric resumed their research in the fall.

Lately they, too, had been using beams of alpha particles from polonium to probe the atomic nucleus. The Curie lab's enviable and constantly replenished stockpile of polonium put them at an advantage over most other researchers. When they decided to repeat and extend the recent experiments of Walther Bothe and Herbert Becker, they began by preparing a polonium source of ninety-eight millicuries— far more intense than the seven-millicurie source that the German researchers had employed. They channeled its diffuse emission of alpha particles into a narrow, collimated beam and aimed it alternately at targets of boron, beryllium, and lithium, to see whether these nonradioactive elements could really be made to release radiation.

In each case, copious radiation streamed from the target into the ionization chamber, where its strength was registered on an electrom- eter. They tried deflecting the emission with magnets but could not. Since it proved neutral—neither positively nor negatively charged— they took it for gamma radiation. They placed thin metal screens in the radiation's path, and it punched right through them. When they replaced the metal screens with others made of hydrogen-rich material such as paraffin, the ionization actually increased in strength.

How to explain this result? It was surprising enough to elicit gamma radiation from lightweight elements. Now it seemed that the gamma radiation was actually knocking hydrogen nuclei out of the paraffin screen, and these ions of hydrogen were adding their power of ionization to that of the gamma rays from the targets.

Irène and Frédéric prevailed on physicist and Curie family friend Jean Perrin to deliver their news to the *Académie des Sciences*.

Chapter Twenty-Eight

WILLY (Beryllium)

———

"I DON'T BELIEVE IT," Ernest Rutherford told his Cavendish crew upon reading the report in the *Comptes rendus* by the team of Curie and Joliot. Of course he trusted the couple to tell the truth. He simply could not come to their conclusions. He felt certain they had misinterpreted their data.

James Chadwick, who was once Rutherford's student at Manchester and now his colleague at Cambridge, took up the problem. Repeating the experiments tried in Giessen and in Paris, Chadwick, too, detected a super-penetrating emission from bombarded targets of beryllium and boron. But it was not gamma radiation. Instead, the powerful emission from the struck targets consisted of never-before-seen neutral nuclear particles—entities originally proposed by Rutherford in 1920.

These "neutrons," Chadwick argued, had been dislodged from the nuclei of the light-element targets. As neutral particles, they avoided deflection by magnets. And since a neutron about equaled the size of a hydrogen nucleus, a neutron could readily knock a hydrogen ion out of a paraffin film.

The existence of neutrons did more than resolve the curious results of recent studies. All up and down the periodic table, neutrons proved the missing link between atomic number and atomic weight. The light target element beryllium, for example, at number four on the table, owed its established atomic weight, nine, to the presence of five neutrons in addition to its four positively charged particles, or protons. Neutrons thus supplied the long-sought "neutral ballast" needed to account for the weight differences between isotopes of any given element.

Among radioelements, neutrons solved the riddle of beta decay. Each neutron presumably consisted of a proton and an electron bound together, so that the breaking of that bond would free one electron for release from the parent element and, at the same time, add one unit of positive charge to the daughter product.

Irène and Frédéric had missed out on a major discovery. Despite their disappointment, they managed to see poetic justice in the revelation of Rutherford's proposed neutral particle by Rutherford's protégé, at Rutherford's lab. As Frédéric later conceded in a lengthy interview, "the genius Rutherford" had presciently conceived "a hypothetical neutral particle which, together with protons, made up the nucleus." Though it seemed far-fetched at the time, the idea had permeated "the atmosphere of the Cavendish Laboratory where Chadwick worked, and it was natural—and just—that the final discovery of the neutron should have been made there."

Warming to his theme, Frédéric went on, "Old laboratories with long traditions always have hidden riches. Ideas expressed in days past by our teachers, living or dead, taken up a score of times and then forgotten, consciously or unconsciously penetrate the thought of the workers in these laboratories and, from time to time, they bear fruit: that is discovery."

He and Irène consoled themselves that their day would come, facilitated by the "hidden riches" of the Curie lab. For now they welcomed the timely arrival of Pierre Joliot, born March 12, 1932, and described in the private journal of his mother as a plump baby with a little bit of chestnut-colored hair, a turned-up nose, and a wise air about him.

Irène, who had languished two weeks in bed following her first *accouchement*, and taken frequent rest periods necessitated by anemia, rebounded with surprising speed. Barely a month after giving birth, she felt well enough to attempt a strenuous new scientific pursuit. Leaving *la petite* Hélène and the infant Pierre behind, she and Frédéric set off for a research station perched on the Jungfraujoch in the Swiss Alps to investigate "cosmic rays." Mysterious radiation from outer space was ionizing the upper reaches of the atmosphere—stripping electrons off air molecules and generating cascades of charged particles. This situation had been discovered with the help of electrometers carried first to the top of the Eiffel Tower and then to greater heights on balloon flights. Scientists, particularly radioactivists, were monitoring the activity from locations all over the world.

"We have already done quite a lot," Irène wrote Marie on May 1, after a week of measuring ionization currents under various weather conditions. "Fred works with the Pohl electrometer and I with the Wulf. This way we can tell the real variations from random signals in one of the devices." She compared their surroundings to "a magnificent nest of glaciers." For once she was not sorry that Marie could not join them. The very thin air above eleven thousand feet caused them headaches and insomnia, she said, while the extreme cold kept them bundled in their ski clothes round the clock. Plus, only one walking path was accessible at this time of year.

Frédéric spared his own mother the worrisome details. "It's a dream to work so comfortably in this fine Institute," he wrote her, praising the view, the food, and the occasional opportunities for skiing on the glaciers. "However, we are sorry not to be with you, and to be deprived of Hélène and Pierre."

———

WHILE IRÈNE and Frédéric hunted cosmic rays, Marie, now sixty-four, was walking with a neighbor on a familiar trail in Cavalaire when she tripped and fell and broke her arm. She developed a fever on top of the injury, and the combination kept her stranded, under medical care, for two weeks beyond the scheduled end of her brief vacation. "I shudder to think of the disruption caused by my absurd accident," she fretted to Irène. "What will I find when I return to Paris!" She implored Irène "to look after Mlle. Lub"—Wilhelmina Lub, due any day from the University of Amsterdam—"to avoid giving her the impression of disarray at the very start of her stay."

"Willy" Lub, the newest researcher, defended her doctoral thesis in Amsterdam on May 4, 1932, and presented herself at the Curie Lab on the tenth. She came with the special endorsement of her mentor, Pieter Zeeman, who deemed her one of his best students in forty years of teaching. Zeeman had shared the 1902 Nobel Prize in Physics with his mentor, Hendrik Lorentz, for revealing the effect of magnetic fields on the appearance of spectral lines. He saw in the thirty-two-year-old Willy Lub "a lively intelligence and a great modesty." There were times, he noted, when she went at her research with such zeal that he had intervened to slow her down for fear her health would suffer.

In Paris, Mlle. Lub accepted Madame's challenge to determine the spectrum of actinium. Three decades after André Debierne's discovery of the element, the Curie lab had yet to accumulate enough actinium to settle its atomic weight, visualize its spectrum, and cement its place on the periodic table. Actinium was not only exceedingly rare but also extremely difficult to extract from pitchblende by chemical means.

Before this project could begin in earnest, and notwithstanding her broken arm, Marie went to Warsaw at the end of May with Dr. Claudius Regaud for the opening ceremonies at the Marie Sklodowska-Curie Radium Institute and Hospital. She even posed for a picture wielding a shovel, seemingly planting a tree with Dr. Regaud in front of the new building while Dr. Bronya Sklodowska-Dluska and the current president of Poland, chemist Ignacy Mościcki, looked on.

Bronya now filled her late husband's intended place as director of the institute. As someone who managed her grief by keeping busy, she ran scant risk of having too little to do. Marie, too, knew the solace—and the strain—of stepping into a deceased loved one's shoes. "Even though you feel lonely," she told Bronya, "you have one consolation: there are three of you in

Mme. Curie and Dr. Claudius Regaud at
the Radium Institute in Warsaw

Warsaw, for companionship and protection. Believe me, family solidarity is what matters most. I was deprived of it, so I know." Of course Marie's siblings had rushed to her side when Pierre died, but then, of necessity, they had gone home. "Take comfort from the ones close, and don't forget your Parisian sister. Let's see each other as often as possible."

Curiethérapie for cancerous tumors had begun months ahead of the hospital's official opening, using the gram of radium purchased in 1929 with the US gift of $50,000. Marie had seen the cost of radium cut in half between her two trips to America, and new veins of high-grade uranium ore found recently near Canada's Great Bear Lake promised another price reduction. True to her established practice of courting reliable radium sources, Marie was already negotiating with mine owners in the Northwest Territories.

A cheaper supply of radium to meet the medical demand would also, alas, aid the purveyors of dangerous, unregulated nostrums such as the radium-containing Radithor. American steel magnate Eben Byers, who boasted that he drank three bottles of Radithor per day to boost his vitality and sexual prowess, suffered a gruesome death by radium poisoning in 1932.

In France, the Tho-Radia product line touted "a scientific approach to beauty" in glowing print advertisements for radioactivity-enhanced cold cream and face powder that purportedly erased signs of aging. Worse, Tho-Radia's claims were backed by the fictitious authority of "Dr. Alfred Curie." Ève sought legal counsel on Marie's behalf, in defense of the family name. To confront the bogus "Dr. Curie," the real Mme. Curie was advised to compose a clarifying statement on *Laboratoire Curie* stationery—using her own words but incorporating the phrases provided as a guide—then sign the document and send it to the managers of every important publication in the country.

———

AT LEAST PART of the reason Irène and Frédéric had failed to recognize the neutron, they rationalized, was their lack of the latest instrumentation. James Chadwick at the Cavendish had augmented his ionization chamber with an amplifier and oscillograph that helped him perceive the particulate nature of the radiation from the bombarded targets. The Joliot-Curie couple, unwilling to be outpaced again by anyone, set about upgrading their ionization chambers, electrometers, and other equipment.

Since the days of her doctoral research, Irène had often relied on devices called "cloud chambers," invented by Scottish meteorologist C. T. R. Wilson, for tracking the paths of subatomic particles. In the same way that real clouds formed from the condensation of water vapor around dust particles, the moist air inside a cloud chamber collected around ions. Particles shooting through the chamber stripped electrons off the gas molecules they passed, ionizing them. Each of these ions seeded a tiny cloud, and together the cloud puffs marked the intruders' paths. A magnetic field around the chamber caused the paths to veer one way or the other in the case of charged-particle projectiles. An observer could photograph the telltale trails through a small glass window in the chamber.

Frédéric devised a variable-pressure cloud chamber. By reducing the pressure inside to extremely low levels—approaching one-hundredth that of the air in the lab—he could lengthen the distances that particles traveled, making the images easier to decipher.

In mid-August that summer, just six months after the debut of the neutron, observations with a cloud chamber brought a second new subatomic particle to light. This happened not in the Curie lab but at the California Institute of Technology, where American physicist Carl Anderson captured compelling evidence for an anti-electron—an electron with positive charge. Anderson called his electron-sized, positively charged particle a "positron."

In the wake of that announcement, Irène and Frédéric reviewed several of their own cloud-chamber photographs, where they had seen what looked like electrons moving in the wrong direction through a magnetic field. Early in 1933, with considerable remorse, they reclassified these errant electrons as positrons.

For the second time in the space of one year, a discovery within their reach had eluded their grasp.

Chapter Twenty-Nine

MARIE-HENRIETTE, MARIETTA, et al.

(Aluminum)

———

MARIE HAD EVERYTHING Pierre had ever wanted. By 1933 her large laboratory employed more than sixty scientists, her course at the university turned new students into radioactivists each year, and the factory at Arcueil made the radioelements needed for research.

She had feared, when Pierre died, that Irène might falter without his guidance. Instead, their daughter had found her own place here in the lab.

Closing in at last on the evasive element actinium, Marie was aided by a bevy of associates that included her longtime colleague Catherine Chamié, who still taught school in the mornings at the *Lycée russe*; the newcomer Willy Lub; Sorbonne graduate student Marie-Henriette Wibratte; and a superb lab technician named Marguerite Perey. In April 1933, a highly regarded radioactivist from the Vienna Radium Institute, Marietta Blau, also joined the group as a visiting scholar on a grant from the International Federation of University Women.

With the same openness that Marie had shown years earlier in regard to her hard labor on radium, she published the Curie lab's step-by-step procedure for obtaining actinium. The lengthy, laborious recipe, years in the making, began with a slurry of Congolese pitchblende waste dissolved in water and hydrochloric acid, to which various precipitants, neutralizers, oxides, nitrates, and other reactants were added one at a time, and the matrix subjected to dozens of precipitations and fractionations. "The treatment, on the whole, presented numerous complications," she explained in the *Journal de Chimie Physique*, "because the chosen chemical operations frequently proved incomplete and had to be repeated, either partially or totally."

As an alternate plan, she suggested perfecting the extraction of actinium's parent element, proto-actinium, and then waiting for it to produce offspring. The process would perforce be slow, given proto-actinium's half-life of approximately thirty thousand years, yet two decades' time might yield as much as half a milligram of actinium—ten times what she had gleaned—with a lot less work.

When Bronya came to visit at Easter time, Marie took her on a scenic drive several hundred kilometers south to Luz, on the edge of the Pyrenees near Lourdes, at a place Irène had chosen as the latest site for replenishing her red blood cells. "We had no trouble with the car," Marie wrote to Ève, whose Citroën she had borrowed, "except it doesn't do all that well on the hills." She declared herself pleased to be in the Pyrenees and not the Alps, for hiking as well as driving, since these lower heights were better suited to "my possibilities."

"I wish you a good rest, and a long one," Ève replied. "You were very tired when you left, and your face looked drawn." Marie still felt tired a week later when she and Bronya returned to Paris. She had counted on a second road trip with her sister, this time to Cavalaire, but the League of Nations pressured her at the last minute to drop everything and preside over a conference on "The Future of Culture," to be held the first week of May in Madrid. When Marie consented to go, Bronya saw her off but stayed in Paris near her nieces.

Marie had to admit that her years of service with the League's International Committee on Intellectual Cooperation had borne very little fruit. None of her pet projects, such as sponsoring science students to spend their summer vacations in foreign laboratories, had been realized. Even her major goal of creating an annotated international bibliography of research in specific disciplines was stalled for lack of enthusiasm: The relevant government agencies, publishers, and professional societies seemed unwilling to relinquish their own traditions, customs, and tastes for the greater good.

As a rule Marie favored small programs of direct benefit to some few over the sort of lavish public event that "The Future of Culture" promised to be. Still, the League's summons to Madrid gave her a platform for speaking about science to an audience of mostly nonscientists. "I stand among those who think that science has great beauty," she said in her address to the congress. "A scientist in the laboratory is not only a technician, but also a child confronted by natural phenomena more enchanting than any fairy tale."

To be a scientist, she maintained, was to follow one's curiosity wherever it led—"to push against the limits of knowledge, to pursue the secrets of matter and of life with no preconceived idea of the eventual outcome."

———

PAUL LANGEVIN, reputed to be the foremost physical theorist in France, seemed perfectly suited to take over the leadership of the Solvay Physics Councils in 1930, after the death of Hendrik Lorentz. The years had proven Paul to be diplomatic, quick-witted, truly fluent in all three languages of science (French, German, and English), and fully conversant with the new realm of nuclear physics that radioactivity, relativity, and quantum theory had revealed.

Paul's longstanding marriage to Jeanne Desfosses, unfortunately, held no more happiness now than before. Outside its bounds, he formed a relationship with his former student Éliane Montel, who had entered the Radium Institute on the strength of his recommendation in 1927, and remained there till she joined his lab at the *Collège de France* in January 1931. They had a child together, Paul-Gilbert Langevin, born July 5, 1933.

No public outcry attended this infidelity, or the illegitimate birth. Nor did any challenge to Paul's authority arise as he changed some established Solvay Council protocol in preparation for the seventh session coming up that October in Brussels. He began by increasing the traditionally small number of invitees to forty, while asking only a handful of them to prepare formal remarks. He was setting the stage for extensive discussion around the few presentations on this year's theme, "Structure and Properties of the Atomic Nucleus." In another innovation, Paul pressed the several designated speakers to submit their written reports in September, allowing time for these to be translated and distributed to the entire group for reading beforehand.

Paul tapped James Chadwick of the Cavendish Laboratory to elaborate on the discovery of the neutron, and the team of Irène Curie and Frédéric Joliot to describe their experiences with "penetrating radiation of atoms under the influence of alpha rays."

Frédéric arrived in Brussels fresh from physics conferences in Leningrad and Moscow. Irène, only the second female physicist ever admitted to the closed circle of a Solvay Council, encountered a third one there, Lise Meitner. Although Professor Meitner had lately been stripped of her university teaching privileges by harsh new racial laws enacted under Chancellor Adolf Hitler, she still headed the physics section at the Kaiser Wilhelm Institute for Chemistry in Berlin.

Marie's friend Albert Einstein did not attend the 1933 Council. Horrified by Hitler's rise, Einstein had renounced his German citizenship in March and was actively seeking a new permanent residence.

"We have with us young people of the very highest caliber from all over Europe and America," Paul Langevin announced on the first day of discussions. As the only participant other than Mme. Curie to have attended every single council, Langevin declared himself pleased indeed to see the infusion of youth. "Young physics needs young physicists," he said. "We are counting on the young ones; it is they who will be writing the papers and carrying out the greater part of the work."

Throughout the preceding year, Irène and Frédéric had crashed their alpha particles into a wide array of substances, followed the ensuing neutrons, and watched positrons make tracks through cloud chambers. One notable observation of theirs concerned the dual behavior of aluminum under bombardment by alpha particles. With a direct hit, an atom of aluminum absorbed the alpha particle, now known to consist of two protons and two neutrons. The aluminum

Seventh Solvay Council, Brussels, 1933: Irène Curie is seated second from left, with her husband immediately behind her and Niels Bohr to her left. Mme. Curie sits near the center with Chairman Paul Langevin to her left. Lise Meitner is seated second from right, with James Chadwick to her left and Ernest Rutherford four places to her right.

atom transformed into an atom of silicon, setting one of the protons free. But sometimes—and this was the big surprise—sometimes the collision yielded a silicon atom, a neutron, and a positron. Irène and Frédéric called these positrons "the positrons of transmutation."

Lise Meitner, who had undertaken similar experiments, spoke up first to object to the French team's idea. "I have not observed the positrons your hypothesis implies," she said. Other council members also raised skeptical questions, leading to a debate.

"We were fairly desolate at that point," Frédéric said later. But during a break Niels Bohr, the reigning architect of atomic structure, approached the couple. "He took my wife and me aside to tell us that he thought our results very important." A bit more encouragement came from Irène's childhood friend Francis Perrin, now a professor at the *Collège de France*, and from Wolfgang Pauli of the Swiss Federal Institute of Technology in Zurich, the author of an influential new textbook on quantum physics. Others seemed to think Irène and Frédéric had blundered or missed something yet again.

Stung by the coolness of their reception in Brussels, they turned to other lab work over the next few weeks. At the end of November, a note reached them from Lise Meitner, saying she had retried her own attempts and refigured the statistics, and now her findings actually did align with theirs.

In January 1934, acting on a suggestion from Francis Perrin, the couple rejigged their experiment and worked at a frantic pace over the next few days to arrive at a striking conclusion.

Chapter Thirty

ÈVE

(Radiophosphorus)

"THE JOLIOTS have completed important new work on a novel phenomenon they discovered in radioactivity," Marie wrote right away to Ève, who was vacationing in Switzerland. "They are on a good path to great success, and they have surely earned it through tremendous effort. I'll explain what they did when you get back, as it's too much to describe in a letter."

The first time through, Irène and Frédéric had skipped a step. In their presentation to the Solvay Council, they claimed that alpha-bombarded aluminum, element number 13 on the periodic table, had transformed directly into silicon, element number 14, with the release of a proton or of a neutron and a positron. But as the couple came to realize in retrials, the changes happened in two phases. First the aluminum nucleus, by absorbing the alpha particle, transmuted to phosphorus, element number 15. And an instant later the new phosphorus atom—a previously unknown and radioactive isotope of phosphorus—decayed to silicon.

The unexpected, unstable interim phosphorus isotope proved itself the chemical twin of familiar phosphorus by the way it reacted with a solution of acids and zirconium salts. This assay needed to be performed rapidly, given the brief half-life of radiophosphorus, which Frédéric and Irène determined by the clicks of a Geiger counter to be three minutes and fifteen seconds.

Having captured one new radioisotope, the couple tried to create others in the same way. By bombarding boron with alpha particles, they produced a short-lived new radioactive isotope of nitrogen. Bombarding magnesium yielded radiosilicon—a third new radioisotope never seen in nature. It seemed that their technique would allow them to artificially inseminate almost any element with the awesome power of radioactivity.

The ability to fashion radioisotopes from common materials, instead of extracting traces of rare ones from tons of exotic ores, suggested crucial practical applications. In tumor treatment, for example, lab-crafted radioisotopes could do their assigned work and decay directly to a stable element—without generating a long and dangerous train of radioactive daughter products. In basic research, tailor-made radioactive markers could be attached as tags to essential nutrients or disease organisms to reveal their pathways through the bodies of animals and plants. A myriad of future uses—perhaps a whole new field of physics —might emerge, but here and now, in mid-January 1934, before anyone outside the family had an inkling of what the two of them had done, the thing was beautiful in itself. After six years of collaboration, they had achieved their life-changing, career-defining breakthrough.

Frédéric and Irène presented one of their new radioisotopes to Marie as though it were a golden trophy or a magic ring.

"I can still see her," Frédéric recalled years later, "taking between her fingers, burnt and scarred by radium, the small tube containing the feebly active material. To check what we had told her, she placed it near a Geiger counter to hear the clicks given off by the rate meter. This was without doubt the last great satisfaction of her life."

———

WHENEVER MARIE'S ongoing work with the actinium family kept her at the lab past the dinner hour, she satisfied herself with a piece of bread or a few cookies and a glass of tea warmed on a hotplate, as she had done in her student days. At sixty-six she voiced the same complaints that plagued many people her age—rheumatism in her shoulder, ringing in her ears—and a few that were peculiar to persons in her line of work, such as abnormal blood counts. Sometime soon, with Ève's help and decorating advice, she intended to move out of the great, echoey apartment on the quai de Béthune, into smaller, more modern quarters in a new building near the university. She also wanted to build a little villa on some land she had bought in Sceaux. As she told Bronya, "I feel the need of a house with a garden more and more, and ardently hope that this plan will succeed."

At Easter Marie invited Bronya to Paris, and also to drive with her to Cavalaire, as they had meant to do the previous year—before the League of Nations summoned Marie to Madrid. By the time the sisters reached the vacation house far to the south, Marie felt tired and depressed, and feared an attack of bronchitis coming on. But it was only a cold, and she had Bronya there to nurse her, and the fine weather made the trip worthwhile.

She was still running a slight fever weeks later when she hugged Bronya goodbye in Paris at the Gare du Nord. Even so, she returned to work at the institute, and to the editing of her book, until the late afternoon in May when her persistent fever and increasing chills sent her home to bed.

For the next several weeks Marie roused herself only to make doctor visits. At length Ève suggested a rest at a sanitarium, and selected Sancellemoz in the Haute-Savoie region of southeastern France. The plan was for Ève to accompany her mother on the journey and stay with her at Sancellemoz for most of July. Next the Sklodowski siblings would visit by turns, and then Irène would spend the month of August with her. By autumn Marie would be well.

A sudden worsening of her condition made the train ride to Saint-Gervais a torture for her, and she collapsed upon arrival at the sanitarium. Her fever had reached 104 degrees. Over the next few days, both her red- and white-blood-cell counts plummeted. In lieu of futile blood transfusions, her attending physicians gave her medications to keep her calm and help her sleep. Through the final night of Marie's life, Ève held one of her hands and Dr. Pierre Lowys held the other. The news of her death, at dawn, traveled swiftly around the world.

"Mme. Pierre Curie died at Sancellemoz on July 4, 1934," sanitarium director Dr. François Tobé reported early that morning. "The disease was an aplastic pernicious anemia of rapid, feverish development. The bone marrow did not react, probably because it had been injured by a long accumulation of radiation."

A lengthy obituary appeared in the *New York Times* under the headline MME. CURIE IS DEAD; MARTYR TO SCIENCE. After giving the details of her demise, the article noted that "few persons contributed more to the general welfare of mankind and to the advancement of science than the modest, self-effacing woman whom the world knew as Mme. Curie. Her epoch-making discoveries of polonium and radium, the subsequent honors that were bestowed upon her—she was the only person to receive two Nobel Prizes—and the fortunes that could have been hers had she wanted them did not change her mode of life. She remained a worker in the cause of science, preferring her laboratory to a great social place in the sun . . . And thus she not only conquered great secrets of science but the hearts of the people the world over."

———

IRÈNE AND ÈVE, knowing their mother's wishes, invited only friends and family, including the laboratory family, to the private funeral in Sceaux on Friday, July 6. Marie wanted to be buried as she had lived—simply, near Pierre, and with no undue fanfare. Józef and Bronya came from Warsaw, each bearing a handful of Polish soil to drop into the grave.

In the days that followed, André Debierne quietly succeeded Marie as director of the *Laboratoire Curie*, and Irène assumed his previous role—the role that had originally belonged to Marie—as *chef de travaux*.

The radium standard that Mme. Curie created in 1911 retired from world service at the time of her death. Although the synchronous events were unrelated, both could be laid to radioactivity. In Marie's body, the years of radioactivity exposure—and X-ray exposure—had halted the production of red blood cells. In the sealed glass tube containing the international radium standard, helium and chlorine gas had accu-mulated as a consequence of radioactive decay, till scientists feared the mounting gas pressure might explode the vessel's fragile walls.

The book that Marie struggled to finish before she died, titled sim-ply *Radioactivité*, was completed by Irène and Frédéric and published in two volumes in 1935. That same year Irène became the second woman to win the Nobel Prize in Chemistry.

Irène remembered her earlier visit to Stockholm, as a teenager drafted to stand by her at-once disgraced and honored mother. This time she and her husband bowed together before the Swedish king, accepting the joint award "for their synthesis of new radioactive ele-ments." At the same ceremony, their former rival James Chadwick received the Physics prize for his discovery of the neutron.

Just as Irène and Frédéric had shared the work that led to this stellar recognition, they divided the delivery of the Nobel lecture between

Irène Curie and Frédéric Joliot

them. She spoke first, explaining the physics of what they had done, followed by his account of the chemistry.

Recalling her parents and citing them by name, Irène reminded the Stockholm audience of the "immense consequences in the knowledge of the structure of matter" that radioactivity had wrought. "Nevertheless," she said, "radioactivity remained a property exclusively associated with some thirty substances existing naturally. The artificial creation of radioelements opens a new field to the science of radioactivity, and so provides an extension of the work of Pierre and Marie Curie."

At present, Frédéric pointed out, "we know how to synthesize more than fifty new radioelements—a number already greater than that of the natural radioelements found in the earth's crust. It was indeed a great source of satisfaction for our late teacher, Marie Curie, to have witnessed this lengthening of the list of radioelements, which she had the glory, in company with Pierre Curie, of beginning."

Irène and Frédéric built a house on the lot in Sceaux where Marie had fancied living out her old age as a gardener. Irène designed the home with a huge open room. Here friends, mostly fellow scientists, gathered on Sunday afternoons in a repetition of the salons her parents had hosted long ago.

In 1936 Irène accepted a cabinet position as undersecretary of state for scientific research. She held the rare distinction of being one of only three women selected to serve the newly installed government of the Popular Front—at a time when the women of France did not yet have the right to vote. She agreed to stay at the post for just three months that summer, however, because she believed other scientists were more qualified than she to hold a portfolio, while very few could fulfill her role as a researcher and teacher of radioactivity.

Irène's poor health consigned her to spend the early months of the Second World War at Clairvivre, a tuberculosis sanitarium near Périgueux. She had with her Marie's gram of radium for safekeeping. Hélène and Pierre stayed at l'Arcouest in the care of their Polish cousin Elzbieta Szyler, while Frédéric carried out dangerous underground activities with the Résistance. These included the preparation of

Molotov cocktails and other homemade explosives in his laboratory at the *Collège de France*. In 1944 he managed to get his family safely to Switzerland, and also to help free Paul Langevin from house arrest in Troyes. At the end of the war, Frédéric was made a commander of the Legion of Honor for his military services, awarded the *Croix de Guerre*, and appointed director of the *Centre National de la Recherche Scientifique*. Having opposed the employment of nuclear power in weaponry, Irène and Frédéric promoted its peaceful uses for the rest of their lives.

Irène advanced to the directorship of the *Laboratoire Curie* in 1946, when André Debierne retired. She became known, in her turn, as *la patronne*, but she looked beyond the grounds of the Radium Institute in choosing Orsay as the site of a major new center for nuclear physics research.

In the early 1950s, while the Orsay laboratories were under construction, Irène presented herself repeatedly as a candidate for election to the *Académie des Sciences*. Frédéric had become an *Académicien* in 1946, but despite Irène's scientific merit and her perseverance, she failed to overturn the institution's nearly three-century tradition of barring women.

The planned new Institute of Nuclear Physics at Orsay, about sixteen miles southwest of Paris, still resembled a rough construction site when Irène died in 1956. Its direction, like that of the *Laboratoire Curie*, was tied to the faculty chair of radioactivity that she had held for ten years. Frédéric sought a replacement for her, but there was no one capable or willing—except Frédéric himself. Just as the grieving Mme. Curie had consented, in 1906, to assume the post vacated by her husband, Frédéric took up Irène's title and duties. He found time and strength enough to complete the building project before his own death two years later.

———

ÈVE CURIE APPROACHED the duty of writing her mother's biography with the same dogged, solitary dedication that had seen her through years of patient piano practice. The memory of reading

Colette's *Sido* challenged Ève to achieve a portrait of her own mother that would touch others' hearts. "There are, in the life of Marie Curie," her book begins, "so many great moments that one is tempted to tell her story as a legend."

Madame Curie, published in 1938 in French (1939 in English), established Ève's reputation as a writer. The book became a bestseller, won a National Book Award in the United States, and inspired a Hollywood film, starring the glamorous Greer Garson as Marie.

Ève, who had been too young to follow her mother and sister to the Western Front, left Nazi-occupied France during World War II to join the Free French Forces in England under General Charles de Gaulle. The Vichy government revoked her citizenship and confiscated her property, including Marie's vacation home in Cavalaire.

In November 1941, as a regular contributor to the *New York Herald Tribune*, Ève embarked on a globe-circling journey to report on wartime conditions. She got close to the front lines in Libya, Burma, and the Soviet Union. In Iran, she met the newly installed Shah Mohammad Reza Pahlavi and the prime minister of Poland in exile, Wladyslaw Sikorski. In India she interviewed Jawaharlal Nehru and Mahatma Gandhi; in China, Chiang Kai-shek and Zhou Enlai. Early on, the Japanese bombing of Pearl Harbor nixed the Pacific leg of her itinerary, so she doubled back to New York five months and forty thousand miles later via South America, instead of flying from Singapore to San Francisco as planned.

She collected her articles in a second book, *Journey Among Warriors*, which was published in the United States and England in 1943.

Ève Curie

Returning to Europe, Ève volunteered with the women's medical corps for the Italian Campaign. In August 1944, having won promotion to the rank of lieutenant in the Free French First Armored Division, she participated in the liberation of Paris and was decorated with the *Croix de Guerre*.

While working in the early 1950s as special advisor to Hastings Lionel Ismay, the first secretary general of NATO, Ève met and married American diplomat Henry Richardson Labouisse Jr., and became a US citizen in 1958. For decades she had joked about being the only member of her family never to win a Nobel Prize. Her self-deprecating quip resounded in 1965, when she accompanied her husband to Oslo to accept the Nobel Peace Prize in his role as executive director of UNICEF, the United Nations International Children's Emergency Fund. Over the next fifteen years, Ève, who had no children of her own, traveled alongside Henry as unofficial "First Lady of UNICEF" to aid mothers and babies in scores of countries.

———

IN 1995, THE REMAINS of Marie and Pierre Curie were exhumed and transferred to an honored place in the *Panthéon*, the great domed shrine where French heroes such as Voltaire and Rousseau are interred. As had often been the case during her lifetime, Mme. Curie was the first woman accorded this ultimate tribute.

Ève, a widow of ninety at the time of the *Panthéon* ceremony, flew from her home in New York to take part in the public procession, beside French president François Mitterrand and Polish president Lech Walesa.

One wonders what Marie herself would have thought of her apotheosis, given that she forbade pomp at her own funeral, and considering how carefully she had curated the placement of bodies in the family burial plot—to the extent of raising Pierre's coffin after Dr. Eugène Curie died, so that the husbands and wives could be reunited.

The veneration of Mme. Curie's mortal remains led to several retellings of her life story. At the end of the twentieth century, documents of hers that had long been deemed too personal, such as her grief journal, or too radioactive, such as her lab notebooks, were made available to scholars. By then a new coda capped the great scientist's biography: radium, the element that defined her career—the miracle elixir seen for decades as the cure for cancer—lost its luster due to its undeniable danger. What had once been the Earth's most costly, most sought-after substance was reclassified as toxic waste, and traditional curietherapy yielded to short-lived artificial radionuclides made possible by the work of Irène and Frédéric Joliot-Curie.

Another coda to Marie Curie's story concerns the role she continues to play as iconic female scientist. By some lights she sets a dauntingly high standard—as though success for a woman in science requires at least two major discoveries deserving of two Nobel Prizes. But she herself stood ever ready to explore physics with children, to train young ladies how to teach science to girls, to make X-ray technicians out of women with rudimentary schooling, and to open her laboratory to those who chose to join her in the pursuit of science as a way of life. In this spirit, she inspires followers to seek the happiness she found while pacing near the stove in the old shed, trying to figure out how Nature works.

Epilogue

MARGUERITE

(Francium)

MARGUERITE PEREY joined the Radium Institute in October 1929—not on the recommendation of Paul Langevin or any Sorbonne professor whose opinion Mme. Curie valued, but because she ranked first in her graduating class at the Parisian vocational school for female lab technicians. As a girl growing up in Villemomble, Marguerite had dreamt of becoming a medical doctor. As the fifth child of a widowed mother, however, she settled for a shorter, more affordable education in chemistry. Now that her ability had been recognized, she was rewarded at age twenty with the extraordinary opportunity to acquire the skills pertaining to radiochemistry. Within months, she became Madame's personal assistant and a trusted partner in her work with actinium.

Marguerite mastered the chemical steps necessary to isolate and purify a quantity of actinium sufficient for spectral analysis. Early in 1934 she arrived at a suitable sample—a few one-hundredths of a milligram of actinium in the form of actiniferous lanthanum oxide. As instructed, she carried the prized material to Pieter Zeeman's laboratory in Amsterdam, where the spectroscopy was to be performed. She stayed four months, until Mme. Curie's death interrupted the work. The actinium spectrum was finally achieved three years later by Marguerite's friend and Curie lab alumna, Zeeman's protégée Willy Lub.

Absent her mentor, Marguerite continued providing actinium-family samples to other researchers at the *Laboratoire Curie*. Irène, for example, wanted a concentrated source of actinium in order to specify the element's half-life, thought to be roughly ten years. André Debierne, actinium's discoverer and now director of the lab, was seeking new isotopes. Marguerite, working alone, studied actinium's emissions. The challenge lay in distinguishing actinium's own beta rays from those of its precursors or derivatives. To do so necessitated separating some actinium from its radioactive parents and grand-parents, then hunting quickly, within hours of purification, before the known daughter products popped up and added their betas.

Unexpectedly, Marguerite detected two distinct patterns of beta release. One grew steadily with time and could be attributed to actinium. The other spiked quickly, leveled off, and then dissipated, suggesting an unknown relative with a half-life of about twenty minutes. She concluded that actinium occasionally—maybe once in

Marguerite Perey (left) and Sonia Cotelle, 1930

every hundred decays—released an alpha particle instead of a beta, giving rise to this previously unknown and short-lived radioelement. She labeled the new substance "actinium K"—her candidate for the number-87 spot on the periodic table.

Jean Perrin presented her findings to the *Académie des Sciences* on January 9, 1939. Later Marguerite changed the provisional "actinium K" to the patriotic "francium."

"Everything I have done," she declared, "I owe to Marie Curie."

Marguerite found herself suddenly famous for discovering an element. Egged on by her lab superiors, she pursued a long-deferred university education at the Sorbonne, beginning with the baccalaureate, and defended her doctoral dissertation—on the discovery of francium—in March 1946. She continued her research in the *Pavillon Curie* under the direction of its new *patronne*, Irène Curie, for another three years, until Frédéric Joliot nominated Marguerite for a position with the nuclear research institute opening at Strasbourg.

In 1962, Professor Marguerite Perey, director of the laboratory for nuclear chemistry at the University of Strasbourg, broke with "immutable tradition" by becoming the first woman admitted to the *Académie des Sciences*. A distinction denied to both Marie and Irène Curie now belonged to her alone. And yet, Marguerite's election was more a gesture than a full embrace, since her status as "corresponding member" conferred neither voting rights nor the title of *Académicienne*. Those full honors ultimately accrued, in 1979, to mathematician and physicist Yvonne Choquet-Bruhat, a specialist in general relativity at the Pierre and Marie Curie University in Paris.

———

AFTER MARGUERITE PEREY announced her discovery of francium in 1939, only two blank spaces remained on the periodic table. Element number 85 was isolated in 1940 at the University of California, Berkeley, and aptly named astatine, from the Greek for "unstable." Astatine's longest-lived isotope has a half-life of eight hours. Promethium, element number 61, followed in 1944, at the Clinton Laboratories in Oak Ridge, Tennessee. All thirty-odd isotopes of promethium are radioactive, and most of their half-lives can be measured in minutes.

Since 1940, scientists have fabricated more than twenty new elements heavier than uranium. Many of these "transuranics" bear names honoring heroes in the history of chemistry and physics, from mendelevium and einsteinium to bohrium, rutherfordium, and curium.

The periodic table, first published in 1869, has become an emblem of science displayed in classrooms and lecture halls everywhere, and is often redrawn to make particular points about the elements. It can be—and has been—configured in hundreds of different ways. One version distorts the rigid grid of squares by allotting space on the basis of each element's relative abundance in the Earth's crust. Here the boxes for hydrogen, carbon, and oxygen balloon out, while the spaces for most metals shrink to thin slivers.

Another periodic-table variation color-codes the elements according to their origins over the lifetime of the universe. Hydrogen and helium date back to the first few minutes following the Big Bang, some fourteen billion years ago. Many more elements, such as carbon and oxygen, are forged by nuclear fusion in the cores of long-lived stars like our Sun, while heavier elements, including platinum and gold, erupt from stellar explosions known as supernovae. Uranium and thorium, the progenitors of the polonium and radium discovered in a modest Paris laboratory in 1898, are the fallout from cataclysmic interstellar collisions between massive dying stars.

The Radioactivists

Isabelle Archinard b. 27 February 1903 in Troinex, Switzerland; d. 28 April 1995 at a Catholic old-age home in Carouge, near Geneva.

Antoine-Henri Becquerel b. 15 December 1852 in Paris; d. 25 August 1908 in Le Croisic, of a heart attack.

Marietta Blau b. 29 April 1894 in Vienna; displaced as a Jew during World War II, she escaped to Oslo (thanks to Ellen Gleditsch), to Mexico (aided by Albert Einstein), and then to the United States, where she held numerous positions in industry, on college faculties, and at research facilities, including the Atomic Energy Commission's Brookhaven National Laboratory; d. 27 January 1970 in Vienna, ten years after her return to Europe.

Bertram Borden Boltwood b. 27 July 1870 in Amherst, Massachusetts; d. 15 August 1927 at his summer home in Hancock, Maine, by suicide.

Harriet Brooks b. 2 July 1876 in Exeter, Ontario; m. Frank Pitcher 1907; d. 17 April 1933 in Montreal, after a long but unspecified illness. An obituary written by Ernest Rutherford and published in *Nature* recalled how well known she had been in the years 1901 to 1905 "for her contributions to the then youthful science of radioactivity."

Catherine Chamié b. 13 November 1888 in Odessa; d. 14 July 1950 in Paris, after a thirty-year career of research and metrology at the *Institut du Radium*.

Sonia Slobodkine Cotelle b. 13 June 1896 in Warsaw; d. 18 January 1945 in Paris, of aplastic anemia.

Irène Curie b. 12 September 1897 in Paris; m. Frédéric Joliot 9 October 1926; d. 17 March 1956, in Paris, of radiation-induced leukemia.

Marie Sklodowska Curie b. 7 November 1867 in Warsaw, then part of the Russian Empire; d. 4 July 1934, at Sancellemoz Sanitarium in Haute-Savoie, France, of aplastic anemia.

Maurice Curie b. 12 October 1888 in Paris; m. Raymonde Simonin; became professor of physics at the Sorbonne Institute of Physico-Chemical Biology; d. 2 September 1975 at Bourg-la-Reine.

Pierre Curie b. 15 May 1859 in Paris; d. 19 April 1906 of a traumatic injury on a Paris street.

Jacques Danne b. 1882 in Paris; served as founding editor of *Le Radium*; d. 8 March 1919 in Paris of a sudden illness.

André-Louis Debierne b. 14 July 1874 in Paris; d. 31 August 1949, in Paris, of lung cancer.

Alicja Dorabialska b. 14 October 1897 in Sosnowiec, part of Russian Poland; d. 7 August 1975, in Warsaw, where she is buried near the symbolic memorial she dedicated to colleagues killed in World War II:

> Here the dead lie alive
> in this grave of chemists, whose ashes
> were scattered by the enemy during the years 1939–45
> and did not find a place of silence in a Polish cemetery.

Renée Galabert b. 28 May 1894 in Chartres; headed the Measurement Service in the *Laboratoire Curie* from 1921 to 1933; d. 29 August 1956, in a Paris hospital, of leukemia.

Ellen Gleditsch b. 29 December 1879 in Mandal, Norway; received the first Sorbonne honorary doctorate ever awarded to a woman in 1962, when she was recognized as the "oldest living pioneer of nuclear-physical and chemical research"; d. 5 June 1968, at her country home in Enebakk, of a stroke.

Irén Götz b. 3 April 1889 in Mosonmagyaróvár, a village in the Austro-Hungarian Empire; m. radical activist László Dienes 1913; d. 1941 in Moscow, of typhus contracted during a brief political imprisonment.

Otto Hahn b. 8 March 1879 in Frankfurt; m. Edith Junghans in 1913; d. 28 July 1968, in Göttingen.

Frédéric Joliot b. 19 March 1900 in Paris; m. Irène Curie in 1926; d. 14 August 1958, in Paris, of liver disease attributed to radiation exposure.

Antonia Korvezee b. 8 March 1899 in Wijnaldum, the Netherlands; d. 17 January 1978 in Delft, the city where she taught radioactivity for two decades and was elected the first female professor at its Institute of Technology.

Jeanne Ferrier Lattès b. 6 April 1888 in Montpellier; collaborated with researchers at the *Pavillon Pasteur* on the use of radioelements in cancer detection and earned her doctoral degree in 1926; m. Georges Fournier, her lab partner, in 1929; d. 1979 in Paris.

May Sybil Leslie b. 14 August 1887 in Woodlesford, West Yorkshire; m. chemist Alfred Hamilton-Burr in 1923; d. 3 July 1937 in Bardsley.

Wilhelmina Lub b. 16 March 1900 in Enkhuizen, the Netherlands; determined the spectrum of actinium; d. 23 December 1986 in Enkhuizen.

Stefania Maracineanu b. 18 June 1882 in Bucharest; d. 15 August 1944, of cancer, in her native city.

Branca Edmée Marques b. 14 April 1899 in Lisbon; m. António Sousa Torres 1925; d. 19 July 1986, twenty years after becoming, at age sixty-seven, the first woman in Portugal to gain a tenured professorship on a university science faculty.

Lise Meitner b. 7 November 1878 in Vienna; partnered professionally with chemist Otto Hahn at the Kaiser Wilhelm Institute in Berlin but was forced to emigrate to Sweden in 1938 because of her Jewish heritage; d. 27 October 1968 in Cambridge, England.

Stefan Meyer b. 27 April 1872 in Vienna; m. Emilie Therese Maass in 1910; d. 29 December 1949 at Bad Ischl, Austria, of a heart attack.

Madeleine Monin b. 3 April 1898 in Le Mans; m. Henri Molinier 24 April 1919 and had two daughters; later trained as a nurse, she was cited by the French Red Cross for care of the wounded during World War II; d. 27 November 1976.

Éliane Montel b. 9 October 1898 near Montpellier; never married but had a child in 1933 with Paul Langevin; d. 24 January 1993 in Paris, having outlived her son, a respected musicologist, by six years.

Marguerite Perey b. 19 October 1909 in Villemomble, a suburb of Paris; became the first woman admitted to the *Académie des Sciences*; d. 13 May 1975 at a clinic in Louveciennes, near Versailles, of metastasized cancer.

Angèle Pompéï b. 4 February 1898, on the island of Corsica; d. 13 February 1999 (age 101) in Marseilles.

Eva Julia Ramstedt b. 15 September 1879 in Stockholm; d. there on 11 September 1974.

Erzsébet Róna b. 20 March 1890 in Budapest, where she earned a doctorate in chemistry in 1921; worked with nearly all radioactivity pioneers, including Ellen Gleditsch, who found her a haven at Oslo when Hitler's advance forced Jews out of Vienna; having relocated to the United States in 1941, she worked on the Manhattan Project and at both the Argonne and Oak Ridge National Laboratories; d. 1981 in Oak Ridge, Tennessee, at age ninety-one.

Ernest Rutherford b. 30 August 1871 at Brightwater, near Nelson, New Zealand; m. Mary Georgina Newton of Christchurch in 1900; d. 19 October 1937 in Cambridge, England, following surgery for a strangulated hernia, his ashes interred at Westminster Abbey.

Frederick Soddy b. 2 September 1877 at Eastbourne, Sussex, England; m. Winifred Moller Beilby in 1908; awarded the 1921 Nobel Prize in Chemistry; d. 22 September 1956, at Brighton, broken-hearted after the death of his wife.

Jadwiga Szmidt b. 8 September 1889 in Lodz (then a Polish enclave in the Russian Empire); m. physicist A. A. Tshernyshev in St. Petersburg in 1923; d. April 1940 (exact date unknown) in Leningrad.

Margaret von Wrangell b. 7 January 1877 in Moscow; m. Vladimir Andronikov in 1928; d. 21 March 1932 in Hohenheim, Germany.

Their Families and Associates

Émile Armet de Lisle (head of the *Sels de Radium* factory) b. 28 June 1853 in Nogent-sur-Marne; d. 11 December 1928 in Paris.

Antoine Béclère (Marie's medical colleague and founder of radiology) b. 17 March 1856, Paris; m. Cécile Vieillard-Baron 20 July 1887; d. 24 February 1939 in Paris, of a heart attack.

Eugène Curie (Pierre's father) b. 28 August 1827 in Mulhouse, France; m. Sophie-Claire Depouilly; d. 25 February 1910 in Sceaux.

Jacques Curie (Pierre's brother and Maurice's father) b. 29 October 1855 in Paris; m. Marie Masson; d. 19 February 1941 in Montpellier.

Bronislawa Sklodowska Dluska (Marie's sister) b. 28 March 1865 in Warsaw; m. Kazimierz Dluski in 1890, the year she received her medical degree; d. 15 April 1939 in Warsaw.

Hélène Joliot (Marie's granddaughter) b. 19 September 1927; pursued a distinguished career as a nuclear physicist, Sorbonne professor, and director at the French National Center for Scientific Research; m. Michel Langevin (Paul's grandson), had two children, and is still very much alive at this writing.

Pierre Joliot (Marie's grandson) b. 12 March 1932; m. biologist Anne Gricouroff; became a biophysicist, won election to the *Académie des Sciences,* and held the Chair of Cellular Bioenergetics for twenty years at the *Collège de France*, before gaining his present title of professor emeritus.

Ève Curie Labouisse (daughter of Marie and Pierre) b. 6 December 1904 in Paris; m. Henry Richardson Labouisse Jr. in 1954; d. 22 October 2007 (age 102), at her home in New York City.

Paul Langevin (student of Pierre, colleague and confidant of Marie) b. 23 January 1872 in Paris; arrested for *Résistance* activities during the Second World War; d. 19 December 1946 in Paris and memorialized in the *Panthéon* for his contributions to relativity, quantum mechanics, and sonar.

Hendrik Antoon Lorentz (chair of the first five Solvay Councils) b. 18 July 1853 in Arnhem, the Netherlands; shared the 1902 Nobel Prize in Physics with Pieter Zeeman; d. 4 February 1928, in Haarlem.

Claudius Regaud (Marie's colleague and director of the *Pavillon Pasteur)* b. 30 January 1870 in Lyon; m. pianist Marie Croizet in 1898; d. 28 December 1940 in Couzon-au-Mont-d'Or.

Bronislawa Boguska Sklodowska (Marie's mother) birthdate unknown; d. 9 May 1878, in Warsaw, of tuberculosis.

Józef Sklodowski (Marie's brother) b. 1863 in Warsaw; m. Jadwiga Kamienska; d. 19 October 1937 in Warsaw.

Wladyslaw Sklodowski (Marie's father) b. 20 October 1832 in Kielce, a staunchly Polish city in the Russian Empire; d. 14 May 1902, following gall bladder surgery, in Warsaw.

Zofia Sklodowska (Marie's sister) b. 1 August 1861 in Warsaw; d. of typhus 1 January 1876 at age fourteen.

Helena Sklodowska Szalay (Marie's sister) b. 20 April 1866 in Warsaw; m. Stanislas Szalay in 1896, served as a school inspector and continued teaching till age eighty; d. 6 February 1961.

Annotations

Preface

Only four individuals aside from Marie Curie have been awarded two Nobel Prizes. They are John Bardeen, who won twice for physics, in 1956 and 1972; Linus Pauling, recognized for chemistry in 1954 and for peace in 1962; Frederick Sanger, for chemistry in 1952 and again in 1980; and K. Barry Sharpless, also for chemistry, in 2001 and 2022.

To date, only 5 women can be counted among the 225 Nobel laureates in physics, and only 8 out of 194 chemistry laureates.

Not all the women mentored by Mme. Curie are included in this book. Some, such as Janina Garczynska from Poland, R. Gourvitch of Lithuania, and "Mlle. Larche," darted through the lab too quickly for their careers to be traced in any detail. The best source for capsule biographies of the female members of the Curie lab is *Les femmes du laboratoire de Marie Curie* by Natalie Pigeard-Micault (see Bibliography).

Chapter 1

Before Dmitri Mendeleev published his periodic table in 1869, many other scientists compiled lists and systems aimed at organizing the components of the material world. William Prout suggested in 1816 that all elements were multiples of hydrogen atoms. Systematic approaches were put forward in the nineteenth century by Johann Döbereiner, Leopold Gmelin, Alexandre-Émile Béguyer de Chancourtois, John Newlands, William Odling, and Julius Lothar Meyer, to name a few. Mendeleev's plan triumphed because he did more than anyone else to develop and promote the idea that ordering the elements by their atomic weights revealed periodic repetitions of their properties.

The theories and experiments that led to informed atomic weight designations include the work of Amedeo Avogadro, Jöns Jacob Berzelius, Stanislao Cannizzaro, Henry Cavendish, John Dalton, Joseph Louis Gay-Lussac, Alexander von Humboldt, Joseph-Louis Proust, and William Prout.

The prefix "eka" that Mendeleev used in his provisional names for as yet undiscovered elements means "one" in Sanskrit. "Eka-silicon," for example, appeared one row below silicon, and indeed germanium turned out to fit right there. Although Mendeleev predicted the existence of several new elements, he did not discover any himself.

Chapter 2

The *École Municipale de Physique et de Chimie Industrielles* was founded in 1882 by the city of Paris. Its distinguished alumni include André Debierne, Frédéric Joliot, Paul Langevin, and Georges Urbain. Today, under the new name of *École Supérieure de Physique et de Chimie*, it is a coeducational institution for undergraduate and graduate-level education and research, part of the *Université PSL (Paris Sciences et Lettres)*.

Chapter 3

After Pierre and Jacques Curie discovered the piezoelectric effect in 1880, Gabriel Lippmann posited that the effect should be reversible—not only would certain crystals generate an electric charge when mechanically stressed, but also the application of an electric charge to those same crystals would cause a mechanical change in structure. The Curie brothers proved Lippmann correct in 1881.

Paul Langevin later applied piezoelectricity to the problem of submarine detection by sonar. Today, all sorts of devices, from inkjet printers to quartz watches, from scanning probe microscopes to electric guitars rely on piezoelectricity.

Chapter 4

The science of spectroscopy was born in 1859 when chemist Robert Bunsen and physicist Gustav Kirchhoff, colleagues at Heidelberg, heated known elements to incandescence, passed the light through a slit, a lens, and a prism, and then examined the emission through a magnifier fitted with a scale. They noted that each element produced an individual signature in colors (wavelengths) of light.

The first electroscopes contained a pair of gold "leaves" that spread apart to register the presence of a charged object. By measuring the speed and degree of the leaves' separation, via a scale mounted in the instrument, the observer assigned a quantity to the charge.

Chapter 5

Radium owed its glow—the soft bluish halos that enveloped the flasks and crucibles in the hangar—to the excitation of atoms in the surrounding air by radioactive emissions from radium and radon. Some scientists think the Curies also observed Cerenkov radiation (a phenomenon named for Pavel Cerenkov in the 1930s), which is the visible-light equivalent of a sonic boom, responsible for the blue glow seen around nuclear reactors.

The "induced radioactivity" that contaminated the Curie lab was an accumulation of the products of radioactive decay. Objects in the vicinity of radium did not become radioactive themselves. Rather, the gaseous emanation (radon) wafting through the room transmuted, and coated objects with a residue of solid daughter products.

The many discoveries of Sir Humphry Davy (1778–1829) include the first dental anesthetic (nitrous oxide, or laughing gas) and several elements (potassium, sodium, chlorine). The Davy Medal was first awarded in 1877 to the founders of spectroscopy, Bunsen and Kirchhoff, and in 1882 to Dmitri Mendeleev and Julius Lothar Meyer.

The original 1901 Nobel Prizes were five in number. A sixth prize, in economics, was established by Sweden's central bank in 1968 and is officially called the Sveriges Riksbank Prize in Economic Sciences in Memory of Alfred Nobel.

Chapter 6

Historians wonder what portion of Pierre's bodily ills might have accrued from radiation exposure. Even his fatal accident has been blamed, by some, as much on his faltering gait as on the inclement weather or momentary carelessness while crossing the street.

Chapter 7

The various titles of workers in the Curie lab, from *chef de travaux* to *travailleur libre*, are outlined in "The Research School of Marie Curie in the Paris Faculty, 1907–14," by J. L. Davis of the Unit for the History, Philosophy and Social Relations of Science at the University of Kent, Canterbury, published in *Annals of Science* 52 (1995).

Chapter 8

Sir William Ramsay's discoveries of noble gases presented the first serious challenge to the periodic table. Radioactivity—both the phenomenon of transmutation and the apparent abundance of radioelements—posed the second. The noble gases were originally accommodated in a column at the far-left end of Mendeleev's table. They have since been moved to the far right. The plethora of radioelements remained problematic till 1913, when many were shown to be isotopes, or chemically identical forms of the same element, differing only in atomic weight (and half-life). The existence of isotopes thinned the population of elements and preserved the structure of the periodic table. The understanding of isotopes remained incomplete until the discovery of the neutron in 1932.

Chapter 9

Claude Louis Berthollet (1748–1822), namesake of the street where Ellen Gleditsch lived in Paris, was a French chemist and senator. In Ellen's later years she described his research in a biographical sketch—one of twenty-five such sketches she authored about French scientists, including a book-length treatment of Antoine Laurent Lavoisier, which was published in Oslo in 1956. The rue Cuvier, site of the Curie lab in Ellen's day, honored the great naturalist and zoologist Georges Cuvier (1769–1832).

Actinium B, described as an alpha-particle emitter in Lucie Blanquies's paper, is now known to be the lead isotope Pb211, and to decay by beta emission.

The Hughes Medal, named for British-American inventor David E. Hughes and awarded annually since 1902 for outstanding contributions to the field of energy, went first to J. J. Thomson. Hertha Ayrton remained the only woman to win this recognition for more than a century—until 2020, when Dame Clare Grey, a fellow of Pembroke College, Cambridge, became the second. Other recipients include Alexander Graham Bell, Hans Geiger, and Stephen Hawking.

Chapter 10

The Exhibition of 1851 Scholarship, which brought May Sybil Leslie to the Curie lab, was created from the large surplus of funds raised for the Great Exhibition in the Crystal Palace. The scholarships, which are still awarded today, also enabled the young Ernest Rutherford to leave New Zealand in 1895 for the Cavendish Laboratory, Cambridge.

Chapter 11

The *Académie des Sciences* was established in 1666. By 1910, full membership in the Academy was strictly limited to six Frenchmen in each of eleven sections (geometry, mechanics, astronomy, geography and navigation, physics, chemistry, mineralogy, botany, rural economy, anatomy and zoology, and medicine and surgery), plus two perpetual secretaries and a maximum of twelve foreign associates.

Chapter 12

Ernest Solvay, a wealthy Belgian chemist and industrialist, indulged his scientific curiosity by funding a series of exclusive conferences, at which he could hear the world's foremost scientists debate the latest, most controversial theories and discoveries. The first physics council lasted from October 30 to November 3, 1911. He added a triennial chemistry conference in 1922, at which point Jean Perrin switched his allegiance from the physics council to the chemistry.

Chapter 13

Marie Curie's kidney ailment may have been a manifestation of tuberculosis, which is known to affect the spine, kidneys, and brain, as well as the lungs. If the child Manya Sklodowska inhaled the *Mycobacterium tuberculosis* during her mother's long illness, which seems likely, she could have become infected without developing any symptoms. It is also possible that she transmitted the disease to her daughter Irène.

Chapter 14

Jean Perrin first suggested an atomic structure resembling a miniature solar system in 1901. Hantaro Nagaoka independently proposed a Saturn-with-rings structure in 1903. But J. J. Thomson, who identified the atom's negative particles in 1897, believed they were embedded among their positively charged counterparts. Instead of a ringed planet or a planetary system, he pictured a "plum pudding" model of the atom that prevailed for about a decade, until Rutherford's discovery of concentrated positive charge at the atomic center revived the earlier conception of electrons as exterior orbiters.

Chapter 15

Ionium does not appear on the modern periodic table. Although it was still considered an element at the time of Boltwood's death in 1927, it later proved to be an isotope of thorium.

Zinc sulfide screens, which figured in the discovery of X-rays, provided the first way to visualize alpha particles. Each alpha striking the screen produced a momentary flash of light, or scintillation. These scintillations were counted, with difficulty, by an observer—until the process was automated.

Chapter 17

Nicole Girard-Mangin was apparently mobilized in error, due to some clerical confusion about her name. Nevertheless, the young doctor was more than willing to serve her country. She fought hard to convince the army to let her do so—and to compensate her properly according to her rank.

Nurse Edith Cavell, a native of England who worked in Belgium, cared for wounded soldiers regardless of their national allegiance. She was executed by a German firing squad on October 12, 1915.

Chapter 18

The scientists responsible for the conceptual leap from atomic weight to atomic number include Charles Barkla; William Bragg and his son, Lawrence Bragg; Henry Moseley; and Antonius van den Broek.

In 1949, the International Union of Pure and Applied Chemists elided the given name of element 91, proto-actinium, to protactinium. An intense alpha-emitter, protactinium is one of the rarest elements.

Modern notations for the numerous isotopes of lead are written as the symbol Pb (from the Latin *plumbum*) with the atomic number 82 as a subscript at the lower left and a superscript at right denoting atomic weight. For example, thorium D became $_{82}Pb^{208}$. The atomic weight of ordinary lead is 207. In common parlance among scientists, isotopes are identified simply by their name and atomic weight, such as uranium-235 and uranium-238.

Chapter 19

The International Federation of University Women exists to the present day, though its name has changed to Graduate Women International. In 1984, the Norwegian chapter created a scholarship named in honor of Ellen Gleditsch.

Chapter 20

Dr. Florence R. Sabin, who shared the spotlight with Mme. Curie at Carnegie Hall, was the first female professor at Johns Hopkins, the first woman to join the Rockefeller Institute as a researcher (studying the pathology of tuberculosis), the first woman to serve as president of the American Association of Anatomists, and the first woman elected to membership in the National Academy of Sciences.

The Association to Aid Scientific Research by Women, headquartered in Boston, named its annual prize in honor of Ellen Swallow Richards, the first woman admitted as a full-time student to MIT, where she studied chemistry, established the Women's Laboratory, and served as assistant professor in several disciplines, including industrial chemistry and applied biology.

A plaque at the Standard Chemical Company's Pittsburgh headquarters states that the radium given to Mme. Curie in 1921 was extracted according to a process invented and supervised by Glenn Donald Kammer of the University of Pittsburgh. Kammer, a chemist, later worked in the manufacture of luminous paint, and died of leukemia in 1927.

MacMillan published *Pierre Curie*, translated into English by Charlotte and Vernon Kellogg, in 1923.

Chapter 21

Mesothorium, not found on the modern periodic table, was discovered and named an element by Otto Hahn in 1907. Subsequent research showed that it occurred in two forms: mesothorium 1, with a half-life of about six years, turned out to be an isotope of radium, and mesothorium 2, with a half-life of about six hours, was an isotope of actinium.

Marie's election to the *Académie de Médecine* in 1922 was a singular event; more than twenty years passed before another woman, Lucie Randoin (1885–1960), entered its ranks in 1946.

Albert Einstein's on-again, off-again attitude toward the International Commission on Intellectual Cooperation continued for two years, until he attended the plenary session of July 25, 1924.

Chapter 22

The International Union of Pure and Applied Chemistry, established in 1919, grew out of a conference held in 1860 in Karlsruhe, Germany, where chemists from many countries met to establish international standards for their science. One of their key objectives was a nomenclature system for elements and compounds.

Many textbooks and websites credit German physicist Friedrich Ernst Dorn with the discovery of radon (Rn). Dorn himself, however, never claimed that distinction, which in fact belongs to Ernest Rutherford and the Curies. It was Rutherford who gave the name "emanation" to the gases given off by radio-elements. Thorium emanation, radium emanation, and actinium emanation were considered separate elements—thoron, radon, and actinon—until 1923, when all three proved to be isotopes of radon. Nevertheless, their individual names persisted into the 1940s.

"Undark," the trade name of the self-luminous paint produced by the U.S. Radium Corporation, contained just enough radium to ensure a release of alpha particles, which caused the white pigment zinc sulfide to luminesce.

Chapter 23

Several radioactivists contributed to the "laws of displacement" describing the zigzag course of radioactive decay. Kasimir Fajans, Georg von Hevesy, Alexander Russell, and Frederick Soddy all worked on the problem. Later they argued with each other over credit for the discovery. When asked to reconcile the conflicting claims, Ernest Rutherford said, "I personally feel that the whole question is a very tangled one, for nearly all the people concerned have talked over the matter . . . The consequence is that it is almost impossible without a judge and jury to examine everyone to state the exact origin of the ideas." Rutherford's statement, dated April 2, 1913, and quoted in *Radioactivity* by Marjorie C. Malley, applies equally well to many other theories and discoveries. Rarely can a scientific breakthrough be attributed to a lone individual or an isolated laboratory. Rather, findings build on one another over time and across boundaries.

Chapter 24

Industrialist and philanthropist Ernest Solvay had died in 1922, but his foresight provided for the continuation of the pivotal meetings that bore his name. In May 1912, he created a foundation, the *Institut International de Physique*, to encourage "the researches which would extend and deepen the knowledge of natural phenomena." Two committees, one administrative and the other scientific, directed the foundation's activities. Mme. Curie served on the International Scientific Committee along with Hendrik Lorentz, Ernest Rutherford, Heike Kamerlingh Onnes, and five others. Although Solvay envisioned a thirty-year lifespan for his physics institute, his heirs recognized its value and extended its tenure, so that the gatherings continue to the present day. The most recent one, dedicated to "The Physics of Quantum Information," took place in 2022.

Chapter 25

Francis W. Aston of the Cavendish Laboratory discovered the two-isotope nature of chlorine in 1919. The apparatus he devised to separate isotopes of elements according to their atomic weights, called the mass spectrograph, earned him the 1922 Nobel Prize in Physics.

In the early 1930s, pitchblende deposits from Great Bear Lake in northwestern Canada posed the first serious challenge to ore from Haut-Katanga. Competition between the Belgian and Canadian processors drove the price of radium down to $40,000 per gram by 1938.

Chapter 26

The city of Kristiania changed its name to Oslo in 1925.

Chapter 28

Cosmic rays were discovered in 1912 by Austrian physicist Victor Hess, who ascended in a series of balloon flights and found the air at altitude to be far more highly ionized than at sea level. Cosmic rays are protons and light nuclei originating elsewhere in the Solar System or even beyond it. When the high-speed particles strike the atmosphere, they generate a cascade of effects, including ionization and the creation of particles now known as pions and muons.

The electrometer used by Irène on the Jungfraujoch was devised by German physicist and Jesuit priest Theodor Wulf, one of the first individuals to detect cosmic rays. Frédéric's electrometer was of a type designed by Robert Wichard Pohl of the University of Göttingen.

The invention of the cloud chamber gained Scottish meteorologist C. T. R. Wilson half of the 1927 Nobel Prize in Physics. The other half went to Arthur Holly Compton, in recognition of the "Compton effect," i.e. the lengthening of wavelengths of photons scattered by collisions with electrons.

Chapter 30

Geiger counters, first conceived by Ernest Rutherford and Hans Geiger at Manchester in 1908 and later refined by Geiger's collaboration with Walther Müller, entered the Curie lab in January 1933, when a postdoctoral fellowship brought Wolfgang Gentner to Paris from Frankfurt. The counter registers the ionization of gas atoms, whether by alpha particles or other types of ionizing radiation.

Neither Frédéric nor Irène cared much for the term "artificial radioactivity," as there was nothing artificial about the radioactivity produced. They preferred the more descriptive, more accurate "synthesis of artificial radioelements."

By the time Irène was recognized as a Nobel laureate in chemistry, two Nobel Peace Prizes and three Nobel Prizes in Literature had been awarded to women. Maria Goeppert Mayer became the second female laureate in physics in 1963, and Dorothy Hodgkin the third in chemistry in 1974.

Epilogue

Only four women in history have discovered or codiscovered naturally occurring elements. Marie Curie was first, in 1898, when she intuited and then identified polonium and radium. She was followed by Lise Meitner (protactinium, 1917), Ida Tacke Noddack (rhenium, 1925), and Marguerite Perey (francium, 1939).

Glossary

active deposit – the first solid radioelements resulting from the decay of gaseous emanation (radon).

alpha ray / alpha particle – the positively charged radioactive emission, identical to the nucleus of a helium atom, with two protons and two neutrons.

atom – the smallest entity of an individual element, consisting of at least one proton and one electron.

atomic number – the number of protons in an atom.

atomic weight – the combined number of protons and neutrons in an atom.

beta ray / beta particle – the negatively charged radioactive emission, identical to an electron.

compound – two or more elements chemically bound to form a new material, such as water from hydrogen and oxygen.

cosmic ray – a fast-moving charged particle, accelerated by its travels through space, that arrives at Earth and ionizes air molecules in the atmosphere, generating showers of secondary particles.

curie – a unit of measurement in radioactivity.

curiethérapie / **curietherapy** – the use of radioactivity in medical treatment, primarily for cancer.

cryogenic laboratory – a venue for studying matter at extremely low temperatures.

electrolysis – the process by which a solution is separated into its component ions.

electromagnetic radiation – the full range of light wavelengths, from radio waves through infrared, visible light, ultraviolet, X-rays, and gamma rays.

electron – the negatively charged particle in an atom; the carrier of electric current; the beta emission.

electroscope – a device for assessing electric charge; the earliest ones had a pair of small gold "leaves" that flew apart when a charged object came close.

element – a unique manifestation of matter, ninety-four of which are naturally occurring.

emanation – the original term for radon, the inert gas produced by the radio-active decay of radium, thorium, and actinium.

gamma ray – the electrically neutral and most highly penetrating radioactive emission; a high-energy variation of visible light.

half-life – the time it takes for half of a radioactive sample to decay.

ion – an electrically charged particle.

ionium – a "new element" discovered in 1907 by Bertram Boltwood, later shown to be an isotope of thorium.

ionize – to imbue with electric charge, as by stripping electrons from atoms so that they become positively charged.

isotope – a variant of an element, identical to that element in properties but differing from it in atomic weight.

mesothorium – a "new element" discovered in 1907 by Otto Hahn, later seen as two "new elements," mesothorium 1 and mesothorium 2, which turned out to be isotopes, respectively, of radium and actinium.

meteorology – the study of the atmosphere, especially weather and climate.

metrology – the science and practice of precise measurement.

molecule – the smallest unit of a compound, or of a diatomic gas such as hydrogen, nitrogen, or oxygen.

neutron – a subatomic particle with about the mass of a proton but without charge.

nucleus – the central region of an atom and the locus of its positive charge.

photon – the smallest unit of visible light or other electromagnetic radiation.

positron – the first anti-matter particle discovered; a positive electron.

proton – the positively charged particle in the nucleus of an atom.

quanta – tiny "packets" or "particles" of energy.

quantum mechanics – the theory and tools of mathematical analysis pertaining to subatomic phenomena.

radioactivity – the spontaneous emission of energy and subatomic particles from the nuclei of certain chemical elements.

radiothorium – an isotope of thorium (Th-228) once thought to be a new element.

salt – a soluble, crystalline chemical compound, often made up of an acid and a base, or a metal and a nonmetal.

spectrum – the wavelengths of light emitted by an element when it is heated to incandescence.

The Radioactive Decay Series

1. The Uranium Family

Radioelement and Rays		Half life (years, days, hours, minutes, seconds)
Uranium I		4,500,000,000 y
↓α		
Uranium X₁ (Thorium 234)		24d
↓β		
Uranium X₂ (Protactinium 234)		1.2 m
↓β		
Uranium II (Uranium 234)		240,000y
↓α		
Ionium (Thorium 230)		77,000 y
↓α		
Radium (Radium 226)		1,600 y
↓α		
Radium emanation (Radium 222)		3.8 d
↓α		
Raduium A (Polonium 218)		3.1m
↓α or ↘β		
Radium B	Astatine	27 m, 2 s
(Lead 214)	(Astatine 218)	
↓β	↙α	
Radium C (Bismuth 214)		20 m
↓β or ↘α		
Radium C'	Radium C"	0.00016 s, 1.3 m
(Polonium 214)	(Thallium 210)	
↓α	↙β	
Radium D (Lead 210)		22y
↓β		
Radium E (Bismuth) 210		5.0 d
↓β or ↘α		
Radium F	Thallium	140 d, 4.2 m
(Polonium 210)	(Thallium 206)	
↓α	↙β	
Radium G (Lead 206)		Not radioactive

2. The Actinium Family

Radioelement and Rays		Half life (years, days, hours, minutes, seconds)
Actinouranium (Uranium 235)		710,000,000 y
↓α		
Uranium Y (Thorium 231)		26 h
↓β		
Protactinium (Protactinium 231)		33,000 y
↓α		
Actinium (Actinium 227)		22 y
↓β or ↘α		
Radioactinium (Thorium 227)	Actinium K (Francium 223)	19 d, 22 m
↓α	↙β	
Actinium X (Radium 223)		11 d
↓α		
Actinium emanation (Radon 219)		4.0 s
↓α		
Actinium A (Polonium 215)		0.0018 s
↓α or ↘β		
Actinium B (Lead 211)	Astatine (Astatine 215)	36 m, 0.0001 s
↓β	↙α	
Actinium C (Bismuth 211)		2.1 m
↓β or ↘α		
Actinium C (Polonium 211)	Actinium C" (Thallium 207)	0.005 s, 4.8 m
↓α	↙β	
Actinium D (Lead 207)		Not radioactive

3. The Thorium Family

Radioelement and Rays		Half life (years, days, hours, minutes, seconds)
Thorium (Thorium 232)		14,000,000,000 y
↓α		
Mesothorium I (Radium 228)		5.8 y
↓β		
Mesothorium II (Actinium 228)		6.1 h
↓β		
Radiothorium (Thorium 228)		1.9 y
↓α		
Thorium X (Radium 224)		3.7 d
↓α		
Thorium emanation (Radon 220)		56 s
↓α		
Thorium A (Polonium 216)		0.15 s
↓α or ↘β		
Thorium B	Astatine	11 h, 0.0003 s
(Lead 212)	(Astatine 216)	
↓β	↙α	
Thorium C (Bismuth 212)		61 m
↓β or ↘α		
Thorium C'	Thorium C"	0.0000003 s, 3.1 m
(Polonium 212)	(Thallium 208)	
↓α	↙β	
Thorium D (Lead 208)		Not radioactive

Sources: Samuel Glassone, Sourcebook on Atomic Energy. Princeton, NJ: D. Van Nostrand, 1950.

Argonne National Laboratory, EVS, Human Health Fact Sheet. Argonne National Laboratory (Illinois), 2005.

Marjorie C. Malley, Radioactivity: A History of a Mysterious Science. New York: Oxford University Press, 2011.

Quotation Sources

Chapter 1

Marie's autobiographical essay is included in her biography of Pierre, *Pierre Curie*, published in 1923. All of her reflections quoted in this chapter are taken from that essay.

Excerpts from her letters to her childhood friend Kazia and cousin Henrietta appear in Ève Curie's 1937 biography, *Madame Curie*. Letters from Wladislaw Sklodowski to his daughters are also excerpted in *Madame Curie*.

Chapter 2

The texts of Marie's letters to and from her brother, Józef, are preserved in Ève Curie's biography, *Madame Curie*.

Marie recorded her first impressions of Pierre in *Pierre Curie*.

Fragments of Pierre's diary are held at the *Bibliothèque nationale de France*. This passage is quoted in *Madame Curie*.

Pierre's letter refusing a decoration, also preserved in the *Bibliothèque nationale*, can be read on-line at the Gallica site.

Chapter 3

Marie's letters in Polish to her brother, her father, and her friend Kazia are quoted in Ève's *Madame Curie*.

Marie recorded her impressions of the gorge of the Truyère and the Brittany coast—and the quality of her married life—in *Pierre Curie*.

Her letters to and from Pierre are quoted in *Madame Curie, Pierre Curie, Marie Curie: A Life* by Susan Quinn, and *Marie Curie: Une femme dans son siècle*.

British industrialist and inventor Rookes Evelyn Bell Crompton (1845–1940) made the very positive comments about Marie's work on magnetized steel at a 1904 meeting of the Institution of Electrical Engineers. He was quoted in "Domesticating the Magnet: Secularity, Secrecy, and Permanency as

Epistemic Boundaries in Marie Curie's Early Work," by Graeme Gooday, in *Spontaneous Generations: A Journal for the History and Philosophy of Science* 3, no. 1 (2009) 68–81.

Chapter 4

Marie noted her observations of Irène's development (and later Ève's as well, over a period of fifteen years) in her *cahier des enfants*, preserved at the *Bibliothèque nationale*. It can be read on the Gallica website.

Marie's paper "Rays Emitted by the Compounds of Uranium and Thorium" appears in English in Alfred Romer's *Radiochemistry and the Discovery of Isotopes*, as do two other papers of hers, "On a New Radio-active Substance Contained in Pitchblende," coauthored with Pierre, and "On a New, Strongly Radio-active Substance Contained in Pitchblende," coauthored with Pierre and G. Bémont.

The congratulatory letter about the *Prix Gegner* is preserved at the *Bibliothèque nationale* and can be read on the Gallica website.

Marie's letters in Polish to her sister Bronya are excerpted in *Madame Curie*.

Chapter 5

Marie's assessment of the need to isolate polonium and radium, as well as her descriptions of that work and the "poor, shabby hangar," are taken partly from her biography of Pierre and partly from her doctoral dissertation, which can be read in French or English online.

Recollections from Marie's students at Sèvres are quoted in the book by her star pupil, Eugénie Feytis Cotton, titled *Les Curie et la radioactivité*.

Marie's exchanges in Polish with her father, brother, and sister Bronya are quoted in *Madame Curie*.

Pierre's letter to Henri Poincaré about the 1903 Nobel Prize is quoted in *Pierre Curie Correspondances* by Karin Blanc.

Chapter 6

Marie's letters to her brother in Polish are quoted in *Madame Curie*.

Pierre's correspondence with his friend and fellow physicist Louis Georges Gouy is held at the *Bibliothèque nationale* and quoted in Ève's *Madame Curie*.

Wilhelm Ostwald's assessment of the hangar is quoted in *Marie Curie* by Robert Reid.

Dmitri Mendeleev's complaints about radioactivity are quoted in Michael D. Gordin's *A Well-Ordered Thing: Dmitrii Mendeleev and the Shadow of the Periodic Table*.

Marie's official reply to W. Marckwald, "On the Radioactive Substance 'Polonium,'" was originally published in German in *Physikalische Zeitschrift* 4

(1902–03) and appears in English in *Radiochemistry and the Discovery of Isotopes.*

The text of Pierre's Nobel lecture can be read on the Nobel Prize website.

Marie's journal of her grief is held at the *Bibliothèque nationale* and can be read on the Gallica website.

Chapter 7

Marie's first lecture at the Sorbonne is quoted in *Obsessive Genius* by Barbara Goldsmith.

Harriet Brooks's letters to Ernest Rutherford are held in the Rutherford Collection of Correspondence at the Cambridge University Library.

The conversation between Rutherford and Soddy in the privacy of the McGill lab is recounted in Muriel Howorth's *Pioneer Research on the Atom: Rutherford and Soddy in a Glorious Chapter of Science – The Life Story of Frederick Soddy, M.A., LL.D., F.R.S., Nobel Laureate.*

Harriet's exchange of letters with the Barnard dean is held in the Barnard College Archives and quoted in the biography *Harriet Brooks: Pioneer Nuclear Scientist* by Marelene and Geoffrey Rayner-Canham.

Harriet conveyed her impressions of Mme. Curie in an April 1910 address to the McGill Alumnae Society, which is reprinted in full in the Rayner-Canham biography.

Marie's letter to Eugénie Feytis is held in the archives of the *École Normale Supérieure* and quoted in *Eugénie Cotton (1881–1967): Histoires d'une vie—Histoires d'un siècle* by Loukia Efthymiou.

Rutherford's letters to Prof. Arthur Schuster at Manchester University recommending Harriet Brooks for the John Harling Fellowship are held at the Royal Society and quoted in the Rayner-Canham biography.

Harriet's handwritten letter of resignation from the Curie lab is reproduced as a photocopy in *Les femmes du laboratoire de Marie Curie* by Natalie Pigeard-Micault.

Frank Pitcher's letters to Harriet Brooks were preserved by their son Paul Brooks Pitcher and loaned to Marelene and Geoffrey Rayner-Canham for inclusion in their biography.

Chapter 8

Letters between Marie and her daughter Irène are collected in *Marie / Irène Curie Correspondance: Choix de lettres (1905–1934).* Another collection, *Lettres: Marie Curie et ses filles*, includes letters to and from both Irène and Ève.

The gossip columnist's assessment of the cooperative school is quoted in *Madame Curie.*

Eyvind Bødtker's correspondence is quoted in "Ellen Gleditsch: Pioneer Woman in Radiochemistry" by Annette Lykknes, Helge Kragh, and Lise Kvittingen, published in *Physics in Perspective* 6 (2004); and also in Anne–Marie Weidler Kubanek's biography of Ellen Gleditsch, *Nothing Less Than An Adventure.*

Ellen spoke of being trusted with a fortune's worth of radium at the Curie lab in a 1911 interview with *Urd,* a Norwegian women's magazine, quoted in Lykknes et al.

Ellen reported her observations of Mme. Curie's work ethic in an article she contributed to *Kvinnelige Studenter 1882–1932,* quoted in Kubanek.

Émile Armet de Lisle's letter to Marie, September 21, 1908, is preserved in the *Bibliothèque nationale* and quoted in Soraya Boudia's *Marie Curie et son laboratoire.*

Chapter 9

The entire text of Lucie Blanquies's *Traité de Physique* can be read on the Gallica website, the digital library of the *Bibliothèque nationale.*

Lucie's study, "*Comparaison entre les rayons alpha produits par différentes substances radioactives,*" was published in the *Comptes rendus: CRAS,* T148 (1909): 1753–55.

Ellen commented on the way problems travel from lab to lab in her article for *Kvinnelige Studenter 1882–1932,* quoted in Lykknes et al, "Ellen Gleditsch: Pioneer Woman."

Bertram Boltwood's correspondence with Ernest Rutherford is collected and edited by Lawrence Badash in *Rutherford and Boltwood: Letters on Radioactivity.*

Hertha Ayrton's letter to her daughter, Barbara, is quoted in Evelyn Sharp's *Hertha Ayrton: A Memoir.* Her letter to the *Westminster Gazette* appeared on March 14, 1909.

Madame's letter to Jacques Danne, preserved at the *Bibliothèque nationale,* is quoted in *Marie Curie et son laboratoire.*

Chapter 10

Marie notes Dr. Curie's illness in her *cahier des enfants,* her vigil with him in *Pierre Curie.*

May Sybil Leslie's letters to Arthur Smithells are reproduced with the permission of Special Collections, Leeds University Library, Brotherton Collection.

Marie's description of Irène's grief appears in her *cahier des enfants.*

Harriet Brooks's April 1910 address to the McGill Alumnae Society is reprinted in full in *Harriet Brooks: Pioneer Nuclear Scientist* by Marelene and Geoffrey Rayner-Canham.

Marie's letters to Paul Langevin were made public in the November 23, 1911, issue of *l'Oeuvre,* which biographer Susan Quinn described as "a scurrilous weekly" aiming "to publicly disgrace and humiliate Marie Curie." Nevertheless, Quinn's study of the letters as printed convinced her, as she noted in *Marie Curie: A Life,* "that they are in fact genuine excerpts."

Chapter 11

Ernest Rutherford's letter to Bertram Boltwood following the Brussels congress appears in Badash's collection of *Letters on Radioactivity*.

Marie's letter to Paul Langevin is quoted in Quinn.

Jean and Henriette Perrin each prepared a testimonial account of events related to the Langevin marital difficulties. These were held at the school of physics and chemistry, and restricted until 1990, when Quinn saw and quoted from them in *Marie Curie: A Life*.

Chapter 12

Marie's thoughts on Ève's musical talent appear in her *cahier des enfants*.

Sybil's June 8, 1911, letter regarding "too much radioactivity" is quoted with the permission of Special Collections, Leeds University Library, Brotherton Collection, MS 416/1a.

Irène's letter home from Zakopane appears in *Correspondance*, Ève's in *Lettres*.

Ellen declared chemistry her "everything" in an interview published in *Urd* on January 14, 1911.

H. A. Lorentz's remarks at the first Solvay Council are quoted in *The Solvay Councils and the Birth of Modern Physics*.

Jacques Curie sent his defense of Marie, dated November 9, 1911, and quoted in Quinn, to the director of the school of physics and chemistry for submission to a newspaper.

Gustave Téry's remarks in *l'Oeuvre* are quoted in Robert Reid's *Marie Curie*.

Svante Arrhenius's December 1, 1911, letter to Mme. Curie is held at the Mittag-Leffler Institute and quoted in Quinn. Marie's reply, held at the Royal Swedish Academy of Sciences Center for History of Science, is also quoted in Quinn.

Marie's 1911 acceptance speech can be read on the Nobel Prize website.

Chapter 13

Ernest Rutherford's letters to Bertram Boltwood regarding his opinion of Mme. Curie and his trip to Paris for the comparison of radium standards appear in the Badash collection.

Marie's "silkworm" letter to her niece Hanna is quoted in the French edition of Ève's *Madame Curie* but excluded (perhaps accidentally?) from Doubleday's English edition.

The 1912 correspondence between Hertha Ayrton and Marie Curie appears in Evelyn Sharp's *Hertha Ayrton: A Memoir*.

The letter inviting Mme. Curie to head a radium institute in Poland is reproduced in Ève's *Madame Curie*.

"Mme. Sklodowska's" letters to Ellen Gleditsch in the summer of 1912 are held at the National Library of Norway and can be viewed on its website.

Hertha's regrets about being too politically active are quoted in Sharp.

Irén Götz's goodbye letter is reproduced as a photocopy in *Les femmes du laboratoire de Marie Curie*.

Chapter 14

The report coauthored by Mme. P. Curie and H. Kamerlingh Onnes, "The Radiation of Radium at the Temperature of Liquid Hydrogen," appeared in *Communications from the Laboratory of Physics at the University of Leiden* 135, 1537 (1913), and is reprinted in *Oeuvres de Marie Sklodowska Curie*.

Marie's sympathy note to Ellen, dated May 1, 1913, is held at the National Library of Norway and quoted in Kubanek.

Ève recollected the conversations between her mother and Albert Einstein in *Madame Curie*.

Einstein's letter of November 23, 1911, is published online in English translation by Princeton University Press, *The Collected Papers of Albert Einstein*.

An account of Margaret Todd's coining of the word "isotope" appears in *Women in their Element*.

Chapter 15

These letters between Marie and her daughters appear in *Lettres: Marie Curie et ses filles*.

Boltwood's letter about Ellen Gleditsch, dated September 12, 1913, is reproduced in Badash. His letter to Ellen herself, dated the day before, is quoted in Kubanek.

Ellen recalled and paraphrased Theodore Lyman's rejection in a 1964 interview with *Dagbladet*, quoted in Kubanek.

Boltwood's article "The Origin of Radium" appeared in *Nature* 76, no. 1978, 544–45 (1907).

Chapter 16

Marie's letters to and from her daughters regarding the First World War are collected in *Lettres: Marie Curie et ses filles*.

The wartime disposition of Mme. Curie's radium was stated in a letter to her from cabinet chief Pierre Guesde, held at the *Bibliothèque nationale* and quoted in *Marie Curie: Une femme dans son siècle*.

Danysz's letter from the front is held at the *Institut Curie* and quoted in Quinn.

Maurice's wartime reports to Marie are held at the *Bibliothèque nationale* and quoted in Quinn.

Marie's wartime letter to Paul Langevin is quoted in Ève's *Madame Curie*.

François Canac's February 1915 letter is held at the *Institut Curie* and quoted in Quinn.

Marie reported her experience of planting bulbs while bombs fell in *Pierre Curie*.

Chapter 17

Hertha's description of and comments about the Ayrton fan are quoted in Sharp.

Maurice's letter to Marie is held at the *Bibliothèque nationale* and quoted in Quinn.

Marie reported the difficulty of preparing emanation bulbs in *Pierre Curie*.

Marie's letter of June 22, 1916, asking Ellen to work at *Sels du Radium*, is preserved in the National Library of Norway and quoted in Lykknes et al., "Ellen Gleditsch: Pioneer Woman."

Ève's letter to Marie appears in *Lettres: Marie Curie et ses filles*.

Excerpts from the text by Ellen Gleditsch and Eva Ramstedt appear in "Ellen Gleditsch: Duty and Responsibility in a Research and Teaching Career, 1916–1946," by Annette Lykknes, Lise Kvittingen, and Anne Kristine Børresen, *Historical Studies in the Physical and Biological Sciences* 36, no. 1 (2005): 131–88.

Sir Ernest's May 18, 1917, letter to his wife is quoted in *Rutherford: Being the Life and Letters of the Rt. Hon. Lord Rutherford, O.M.* by A. S. Eve.

Marie's January 27, 1918, letter to Ellen, held in the National Library of Norway, is quoted in Kubanek.

Irène's letter to Marie appears in *Lettres*.

Fernand Holweck's letter is quoted in Quinn.

Eugénie Feytis Cotton's observations on peace appear in her 1967 biography of her husband, and are quoted in Loukia Efthymiou's *Eugénie Cotton (1881–1967)*.

Chapter 18

Marie expressed her revulsion to war in *Pierre Curie*. Her reflection on injustices done to Poland also appears in *Pierre Curie*.

Her report co-authored with Dr. Regaud is quoted in "Marie Curie and the Radium Industry: A Preliminary Sketch" by Xavier Roqué, in *History and Technology*, 13 (1997).

Her December letter to her brother is quoted in Ève Curie's biography.

Stefan Meyer's description of postwar privations in Vienna appears in A. S. Eve's *Rutherford: Being the Life and Letters*.

Ernest Rutherford's June 14, 1919, thank-you note to Ellen Gleditsch is quoted in Kubanek.

Marie's August 1 letter to Ellen is in the National Library of Norway and on its website. She expressed her dream of an ideal lab in *Pierre Curie*.

Her letters to her daughters appear in *Lettres: Marie Curie et ses filles*.

Chapter 19

Ellen Gleditsch's letter describing Randi Holwech is quoted in Kubanek.

Marie spoke of her occasional "discomfort" from handling radium in her auto-biographical sketch.

She extolled radon above radium for cancer treatment in *Pierre Curie*.

Mrs. Meloney recalled her first meeting with Mme. Curie in the introduction she wrote to *Pierre Curie*.

The June 12, 1920, postcard to Marie from Ellen, Eva, and Sybil is quoted in Kubanek.

Ellen's comment about the value of study abroad is quoted in "Appreciated Abroad, Depreciated at Home—The Career of a Radiochemist in Norway: Ellen Gleditsch (1879–1968)" by Annette Lykknes, Lise Kvittingen, and Anne Kristine Børresen, *Isis* 95 (2004): 576–609.

Chapter 20

Marie's letter to her daughters from Brussels appears in *Lettres: Marie Curie et ses filles*.

Her letter to Henriette Perrin is quoted in *Madame Curie*.

Ève recorded her memories of the trip to America in *Madame Curie*.

Marie praised American women's colleges for their emphasis on sports in *Pierre Curie*. Her disgruntled letter to the rector at the Sorbonne is quoted in Quinn.

She described the radium presentation ceremony in *Pierre Curie*.

Bertram Boltwood's account of Mme. Curie's visit to America appears in Badash's collection of *Letters on Radioactivity*.

Marie's brief summary of her life story is quoted in Mrs. Meloney's introduction to *Pierre Curie*.

Chapter 21

Welcoming remarks by Dr. Anatole Chauffard are quoted in *Madame Curie*.

Catherine Chamié's correspondence is archived at the *Institut Curie* and quoted in *Les femmes du laboratoire de Marie Curie*. Her observations of Marie's technique are quoted in *Madame Curie*.

The 1922 letters between Albert Einstein and Marie Curie regarding the ICIC are collected in *Albert Einstein: Correspondances françaises*, by Michel Biezunski, and quoted in Quinn.

Marie's letters to her daughters appear in *Lettres: Marie Curie et ses filles*.

Ève described the décor of the family apartment in *Madame Curie*.

Marie's letters to her sister Bronya are excerpted in *Madame Curie*.

Chapter 22

A copy of the Academy of Medicine report, written by Dr. Regaud, is preserved with Mme. Curie's papers at the *Bibliothèque nationale*.

Ève assessed Marie's parenting style in *Madame Curie*.

Frédéric Joliot's early impressions of Irène Curie are quoted in two biographies, one by his scientific colleague Pierre Biquard and the other by his acquaintance Maurice Goldsmith. (See Bibliography.)

Marie's letters home to her daughters appear in *Lettres*.

Harlan Miner's 1925 letters to Mme. Curie are held at the *Bibliothèque nationale*.

Chapter 23

The February 19, 1926, "Fred" postcard is included in *Lettres: Marie Curie et ses filles*.

Emilie Roederer Joliot's impressions of Marie and Irène Curie were recorded in her diary and quoted in the Goldsmith biography of Frédéric.

Marie's letters to Ève appear in *Lettres*.

Frédéric's letters to Irène are quoted in the Goldsmith biography.

Nobuo Yamada's letters are quoted in "Nobuo Yamada (1896–1927)—Marie Curie's First Japanese Disciple" by Keiko Kawashima of the Nagoya Institute of Technology, published online at https://www.ne.jp/asahi/kaeru/kawashima/yamada/yamada.html.

Chapter 24

Irène and Frédéric's first joint paper was "*Sur le nombre d'ions produits par les rayons alpha du RaC' dans l'air.*" *CRAS*, T186 (1928): 1722–24.

Frédéric's letters are interspersed with Irène's in *Marie / Irène Curie Correspondance*.

Marie's letters to Ève appear in *Lettres*.

Irène's notebook about her children is held by the Curie and Joliot Curie Association, and quoted in *Irène Joliot-Curie* by Louis-Pascal Jacquemond.

Marie's letter to her brother is held at the National Museum, Warsaw, and quoted in Quinn.

Marie's comments to reporters about the dial painters appeared in the *New York Journal*, May 26, 1926.

Chapter 25

An interview with Ellen Gleditsch appeared in the Oakland *Tribune* on April 3, 1929.

Ellen's letter to Marie is held at the National Library, Oslo, and is quoted in Kubanek.

Ellen wrote up her experiences abroad in her memoir "Female Students 1882–1932," quoted in Lykknes et al., "Ellen Gleditsch: Pioneer Woman."

Ellen's comments to her fellow university women in Norway were reported in *Dagbladet* on October 7, 1929, and quoted in Kubanek.

Marie's letter to Ève appears in *Lettres*.

Chapter 26

Marie's letters to and from her daughters appear in *Lettres*.

Ellen's comments at the 1930 jubilee are quoted in Lykknes et al, "Ellen Gleditsch: Pioneer Woman in Radiochemistry."

Frédéric's letter to Marie appears in *Correspondance*.

Chapter 27

Marie's and Ève's letters to Irène from Spain appear in *Lettres*.

Marie's letters to Ève from Geneva, to Irène from Rome, and from Irène in the mountains also appear in *Lettres*.

Chapter 28

Ernest Rutherford's disbelief regarding the Paris findings on gamma radiation was reported by physicist Pierre Radvanyi, a former student of Frédéric Joliot, and science historian Monique Bordry in their book *La radioactivité artificielle et son histoire*.

Frédéric's comments on discovery and the hidden riches of old laboratories appeared in an interview for the *Gazette de Lausanne* of June 29, 1957, and are excerpted in the Goldsmith biography, as are his letters to his mother.

Letters between Irène and Marie appear in *Lettres*.

Pieter Zeeman's letter of recommendation for Willy Lub, dated January 23, 1932, is quoted in *Les femmes du laboratoire de Marie Curie*.

Marie's letter to Bronya is quoted in *Madame Curie*.

Chapter 29

Marie's article "Sur l'Actinium" appeared in the *Journal de Chimie Physique* 27 (1930) and is included in *Oeuvres de Marie Sklodowska Curie*.

Marie's letters to and from Ève appear in *Lettres*.

Marie's remarks at the Madrid conference are quoted in *Les Curies* by Eugénie Feytis Cotton.

Paul Langevin's welcoming remarks were recorded in the official proceedings of the seventh Solvay Council and quoted in *The Solvay Councils and the Birth of Modern Physics*. Lise Meitner's negative comment about the positrons of transmutation is also quoted there.

Frédéric's reaction to the criticism is quoted in Radvanyi's *Les Curie: Pionniers de l'atome*.

Chapter 30

Marie's letter to Ève appears in *Lettres*.

Frédéric's memory of Marie's reaction to the discovery of artificial radioactivity is quoted in Radvanyi and Bordry, *La radioactivité artificielle*.

Marie's letter to Bronya is excerpted in *Madame Curie*.

Dr. Tobé's announcement of Marie's death is quoted in *Madame Curie*.

Marie's obituary appeared in the *New York Times* on July 5, 1934.

Irène's and Frédéric's Nobel Prize lectures ("Artificial Production of Radioactive Elements" and "Chemical Evidence of the Transmutation of Elements") can be read on the Nobel Prize website.

Epilogue

Marguerite Perey's debt to Mme. Curie is quoted in Cotton's *Les Curies*.

Bibliography*

———

Abir-Am, Pnina G., and Dorinda Outram, eds. *Uneasy Careers and Intimate Lives: Women in Science, 1789–1979*. New Brunswick, NJ: Rutgers, 1989.

Augustin, Marion, with Natalie Pigeard-Micault. *Marie Curie: Une femme dans son siècle*. Paris: Gründ, 2017.

Badash, Lawrence, ed. *Rutherford and Boltwood: Letters on Radioactivity*. New Haven, CT: Yale, 1969.

Biquard, Pierre. *Frédéric Joliot-Curie: The Man and His Theories*. Translated by Geoffrey Strachan. New York: Eriksson, 1966.

Blanc, Karin. *Pierre Curie Correspondances*. Saint-Rémy-en-l'Eau: Monelle Hayot, 2009.

Blanquies, L. *Traité de Physique*. 2nd ed. Paris: Libraire d'Éducation Nationale, 1909.

Bordry, Monique, and Soraya Boudia, eds. *Les rayons de la vie: Une histoire des applications médicales des rayons X et de la radioactivité en France 1895–1930*. Paris: Institut Curie, 1998.

Boudia, Soraya. *Marie Curie et son laboratoire*. Paris: Éditions des archives contemporaines, 2001.

Bragg, Sir William. *Concerning the Nature of Things*. London: G. Bell & Sons, 1925.

Brock, William H. *The Norton History of Chemistry*. New York: Norton, 1993.

Chavannes, Isabelle, ed. *Leçons de Marie Curie*. Les Ulis: EDP Sciences, 2003.

Clark, Claudia. *Radium Girls: Women and Industrial Health Reform, 1910–1935*. Chapel Hill: University of North Carolina Press, 1997.

Clark, Ronald W. *Einstein: The Life and Times*. New York: World, 1971.

Cotton, Eugénie. *Les Curie et la radioactivité*. Paris: Pierre Seghers, 1963.

Curie, Ève. *Madame Curie*. Paris: Gallimard, 1938.

———. *Madame Curie: A Biography*. Translated by Vincent Sheean. New York: Doubleday, 1939.

---

*Journal articles (in *Comptes rendus*, *Le Radium*, *Nature*, *Science*, et al.) are not included here for space reasons.

Curie, Mme. Pierre. *La radiologie et la guerre*. Paris: Librairie Félix Alcan, 1921.

———. *Radioactivité*. 2 vols. Paris: Hermann, 1935.

Curie, Mme. Sklodowska. *Radio-Active Substances: Thesis Presented to the Faculté des Sciences de Paris*. London: Chemical News, 1904.

Curie, Marie. *Pierre Curie*. Translated by Charlotte and Vernon Kellogg. New York: MacMillan, 1923.

Curie, Marie, and Irène Curie. *Correspondance: Choix de lettres (1905–1934)*. Paris: Les Éditeurs Français Réunis, 1974.

Curie, Pierre. *Oeuvres de Pierre Curie. Publiées par les soins de la Société Française de Physique*. Paris: Gauthier-Villars, 1908.

del Regato, Juan A., M.D. *Radiation Oncologists: The Unfolding of a Medical Specialty*. Reston, VA: Radiology Centennial, 1993.

Des Jardins, Julie. *The Madame Curie Complex: The Hidden History of Women in Science*. New York: Feminist Press, 2010.

Efthymiou, Loukia. *Eugénie Cotton (1881–1967): Histoires d'une vie—Histoires d'un siècle*. Mauritius: Éditions universitaires européennes, 2019.

Emsley, John. *Nature's Building Blocks: An A–Z Guide to the Elements*. Oxford: University Press, 2001.

Eve, A. S. *Rutherford: Being the Life and Letters of the Rt Hon. Lord Rutherford, O.M.* Cambridge: University Press, 1939.

Gamow, George. *Thirty Years that Shook Physics: The Story of Quantum Theory*. New York: Doubleday, 1966.

Giroud, Françoise. *Marie Curie: A Life*. Translated by Lydia Davis. New York: Holmes & Meier, 1986.

Goldsmith, Barbara. *Obsessive Genius: The Inner World of Marie Curie*. New York: Norton, 2005.

Goldsmith, Maurice. *Frédéric Joliot-Curie: A Biography*. London: Lawrence and Wishart, 1976.

Gordin, Michael D. *A Well-Ordered Thing: Dmitrii Mendeleev and the Shadow of the Periodic Table*. Princeton, NJ: University Press, 2019. Revised edition.

Grimoult, Cédric. *Marie Curie: Génie persécuté*. Paris: Ellipses, 2023.

Harvie, David I. *Deadly Sunshine: The History and Fatal Legacy of Radium*. Gloucestershire, UK: Tempus, 2005.

Jacquemond, Louis-Pascal. *Irène Joliot-Curie: Biographie*. Paris: Odile Jacob, 2014.

Joliot-Curie, Irène. *Marie Curie, Ma Mère*. Paris: Plon, 2022.

Joliot-Curie, Irène, ed. *Oeuvres de Marie Sklodowska Curie / Prace Marii Sklodowskiej-Curie*. Warsaw: *Académie Polonaise des Sciences / Polska Akademia Nauk*, 1954.

Kubanek, Anne-Marie Weidler. *Nothing Less Than an Adventure: Ellen Gleditsch and Her Life in Science*. St. Mary's, Ontario: Crossfield, 2010.

Langevin-Joliot, Hélène, and Monique Bordry, eds. *Lettres: Marie Curie et ses filles*. Paris: Pygmalion, 2011.

Lévy-Leblond, Jean-Marc, ed. *Lettres à Marie Curie*. Vincennes: Thierry Marchaisse, 2020.

Livingston, James D. *Driving Force: The Natural Magic of Magnets*. Cambridge: Harvard University Press, 1996.

Lykknes, Annette. "Ellen Gleditsch: Professor, Radiochemist, and Mentor." PhD diss., Norwegian University of Science and Technology, 2005.

Lykknes, Annette, and Brigitte Van Tiggelen, eds. *Women in their Element: Selected Women's Contributions to the Periodic System*. Singapore: World Scientific, 2019.

Malley, Marjory C. *Radioactivity: A History of a Mysterious Science*. Oxford: University Press, 2011.

Marage, Pierre, and Grégoire Wallenborn, eds. *The Solvay Councils and the Birth of Modern Physics*. Basel: Birkhäuser Verlag, 1999.

Mehra, Jagdish. *The Solvay Conferences on Physics: Aspects of the Development of Physics Since 1911*. Boston: D. Reidel, 1975.

Moore, Kate. *The Radium Girls*. London: Simon & Schuster, 2016.

Nye, Mary Jo. *Before Big Science: The Pursuit of Modern Chemistry and Physics 1800–1940*. Cambridge: Harvard University Press, 1996.

Ogilvie, Marilyn Bailey. *Marie Curie: A Biography*. Santa Barbara: Greenwood, 2004.

Ogilvie, Marilyn Bailey, and Joy Harvey, eds. *The Biographical Dictionary of Women in Science*. 2 vols. New York: Routledge, 2000.

Pasachoff, Naomi. *Marie Curie and the Science of Radioactivity*. New York: Oxford University Press, 1996.

Paul, Harry W. *From Knowledge to Power: The Rise of the Science Empire in France, 1860–1939*. New York: Cambridge University Press, 1985.

Pigeard-Micault, Natalie. *Les femmes du laboratoire de Marie Curie*. Paris: Editions Glyphe, 2013.

Quinn, Susan. *Marie Curie: A Life*. New York: Simon & Schuster, 1995.

Radvanyi, Pierre. *Les Curie: Pionniers de l'atome*. Paris: Belin, 2010.

Radvanyi, Pierre, and Monique Bordry. *La radioactivité artificielle*. Paris: Seuil, 1984.

Rayner-Canham, Marelene F., and Geoffrey W. Rayner-Canham. *A Devotion to Their Science: Pioneer Women of Radioactivity*. Philadelphia: Chemical Heritage Foundation, 1997.

———. *Harriet Brooks: Pioneer Nuclear Scientist*. Montreal: McGill-Queen's University Press, 1992.

———. *Women in Chemistry: Their Changing Roles from Alchemical Times to the Mid-Twentieth Century*. Philadelphia: Chemical Heritage Foundation, 1998.

Reid, Robert. *Marie Curie*. New York: Saturday Review Press, 1974.

Romer, Alfred, ed. *The Discovery of Radioactivity and Transmutation*. New York: Dover, 1964.

———. *Radiochemistry and the Discovery of Isotopes*. New York: Dover, 1970.

Rutherford, Ernest. *Radioactive Substances and Their Radiations*. Cambridge: University Press, 1913.

Sacks, Oliver. *Uncle Tungsten: Memories of a Chemical Boyhood*. New York: Knopf, 2001.

Scerri, Eric R. *The Periodic Table: Its Story and Its Significance*. Oxford: University Press, 2007.

Sharp, Evelyn. *Hertha Ayrton: A Memoir*. London: Edward Arnold, 1926.

Sime, Ruth Lewin. *Lise Meitner: A Life in Physics*. Berkeley: University of California Press, 1996.

Soddy, Frederick. *The Interpretation of Radium and the Structure of the Atom*. 4th ed. London: John Murray, 1920.

Strathern, Paul. *Mendeleyev's Dream: The Quest for the Elements*. London: Hamish Hamilton, 2000.

van Spronsen, J. W. *The Periodic System of Chemical Elements: A History of the First Hundred Years*. Amsterdam: Elsevier, 1969.

Weeks, Mary Elvira. *Discovery of the Elements*. 5th ed. Easton, PA: Journal of Chemical Education, 1945.

Wilson, David. *Rutherford: Simple Genius*. Cambridge: MIT Press, 1983.

Wirtén, Eva Hemmungs. *Making Marie Curie: Intellectual Property and Celebrity Culture in an Age of Information*. Chicago: University of Chicago Press, 2015.

Illustration Credits

———

Credits for the insert images are as follows: Page 1: National Library of Medicine. Page 2, Image A: Library of Congress; Page 2, Image B: UtCon Collection / Alamy Stock Photo. Page 3, Image A: Wikimedia Commons; Page 3, Image B: Musée Curie (coll. ACJC); Page 3, Image C: History and Art Collection / Alamy Stock Photo; Page 3, Image D: Smithsonian Libraries and Archives. Page 4, Image A: Musée Curie (coll. ACJC); Page 4, Image B: Musée Curie (coll. ACJC); Page 4, Image C: Wikimedia Commons; Page 4, Image D: Wikimedia Commons. Page 5, Image A: Courtesy of the National Museum of Nuclear Science & History; Page 5, Image B: Musée Curie (coll. ACJC). Page 6, Image A: Bibliothèque nationale de France. Département des Manuscrits. NAF 28138; Page 6, Image B: Musée Curie (coll. ACJC). Page 7, Image A: Musée Curie (coll. ACJC); Page 7, Image B: Photo Wyndham Edr. Musée Curie (coll. ACJC). Page 8, Image A: Courtesy of Science History Institute; Page 8, Image B: Wikimedia Commons via NASA.

Credits for the images running throughout are as follows: Page 7: Musée Curie (coll. ACJC). Page 10: Musée Curie (coll. ACJC). Page 13: Ann Ronan Picture Library / Heritage Images / Science Photo Library. Page 20: Musée Curie (coll. ACJC). Page 25: Photo Albert Harlingue. Musée Curie (coll. ACJC). Page 46: Musée Curie (coll. ACJC). Page 48: Photo Albert Harlingue. Musée Curie (coll. ACJC). Page 59: Musée Curie (coll. imprimés). Page 63: Musée Curie (coll. ACJC). Page 72: Wikimedia Commons. Page 83: National Library of Norway. Page 100: Wikimedia Commons. Page 108: Henri Manuel / Roger-Viollet / Granger—All rights reserved. Page 120: Wikimedia Commons. Page 131: Jewish Women's Archive. Page 137: Musée Curie (coll. ACJC). Page 142: Musée Curie (coll. ACJC). Page 156: Musée Curie (coll. ACJC). Page 159: Musée Curie (coll. ACJC). Page 162: © Imperial War Museum (FEQ 491). Page 174: Churchill Archives Centre, Cambridge, MTNR 8/4/1. Page 187: Library of Congress, Prints and Photographs Division. Page 189: Library of Congress, Prints and Photographs Division. Page 190: Musée Curie (coll. Institut du radium). Page 199: Photo Albert Harlingue.

Musée Curie (coll. ACJC). Page 206: Musée Curie (coll. ACJC). Page 229: Musée Curie (coll. ACJC). Page 233: National Library of Norway. Page 246: Musée Curie (coll. ACJC). Page 253: Wikimedia Commons. Page 259: Photo Henri Manuel. Musée Curie (coll. ACJC). Page 262: Sueddeutsche Zeitung Photo / Alamy Stock Photo. Page 266: Musée Curie (coll. ACJC).

Appreciation

———

While studying Marie Curie's career and the opportunities she made available to other women, I was repeatedly struck by the helpful alliances she formed with the men in her life, from her father, who believed in her right to an advanced education, and her brother, with whom she maintained a lifelong correspondence, to the "little brother-in-law" who watched over her in Paris, and of course Pierre—also Pierre's father, who enabled her to be a working mother, and the colleagues who befriended her in the lab, on the university faculty, at conferences, and at council meetings. They reminded me of all the good men to whom I am gratefully indebted, beginning with my father and brothers, Sam, Mike, and Steve Sobel; my son, Isaac Klein; my son-in-law, Justin Kobrin; and my many mentors, especially Will Andrewes, John Bethel, Michael Carlisle, Joseph Cherry, Frank Drake, George Gibson, Owen Gingerich, Ralph Kazarian, Art Klein, Jacob Luria, David May, Kit Reed, Carl Sagan, Bill Stockton, Dick Teresi, and Alfonso Triggiani.

For their various and generous aid as facilitators, advisors, or technical experts on this project, I heartily thank Céline Fellag Ariouet, William B. Ashworth Jr., Elisabeth Bouchaud, Marcin Dolecki, Conor Gruber, Sheryl Heller, Roald Hoffman, Renaud Huynh, Edward Landa, Aurélie Lemoine, Annette Lykknes, Doug Offenhartz, Natalie Pigeard-Micault, Susan Quinn, Kristina Ramsland, Marelene and Geoffrey Rayner-Canham, Jessica Reed, Robert Schumacher, Mary Schwager, Brigitte Van Tiggelen, and Jaroslaw Wlodarczyk.

Index

Note: Page numbers in *italics* reference photographs and figures.

Ayrton, William, 92–93
Ayrton Fan, 161–62, *162*

Barling, Gilbert, *142*
Becker, Herbert, 240, 242
Béclère, Antoine, 101, 152, 164, 194, 238, 271
Becquerel, Henri, 35–36, 41, 43, 49–50, 55, 70, 95, 123, 269
Beilby, George, 134, 139
Bémont, Gustave, 42, 43
beta rays/particles, 50, 57, 61, 70, 215, 223, 266–67
Blanquies, Lucie, 89–90, 94, 95, 99, 110, 168
Blau, Marietta, 249, 269
Bødtker, Eyvind, 82–83, 227
Boguski, Józef, 12, 41
Bohr, Niels, 186, 198, 222, 223, 239–40, *253*, 254
Boltwood, Bertram
  background, 90
  birth and death of, 269
  correspondence with Rutherford, 90, 109, 126, 127–28, 132, 142–43, 191
  Gleditsch's work with, 142–43, 145–47, 175
  Marie's meetings with, 191
  research interests of, 90–91, 145–47
Bonnevie, Kristine, 183, 184, 197, 227
Borel, Émile, 122, 235
Borel, Marguerite, 122
Born, Max, 223
Bothe, Walther, 240, 242
Bouchard, Charles, 55, 114
Bouty, Edmond, 53
Branly, Édouard, 113, 114
Brazilian Federation for Female Progress, 217
Brooks, Harriet. (*See also* Pither, Harriet Brooks)
  background, 71–74
  birth and death of, 269

engagements and marriage of, 75–76, 78–79, 104–5, 110, 132, 183, 204
IFUW chapter founded by, 183
photograph of, *72*
research interests of, 74–75, 76–79
on women in research, 75–76, 79, 104–5
Bureau International des Poids et Mesures (BIPM), 133–35, 232
Byers, Eben, 247

Cabrera, Blas, 237
Cailliet, René, 202–3
Caisse Nationale des Sciences, 235
Canac, François, 158
Carlough, Margaret, 208
Carnegie, Andrew, 77
Chadwick, James, 243–44, 247–48, 252, *253*, 259
Chamié, Catherine, 195–96, 198, 206, 238, 249, 269
Chauffard, Anatole, 194, 195
Chavannes, Alice, 81, 96, 116, 129
chlorine, 52, 161–62, 227
Choquet-Bruhat, Yvonne, 267
cloud chambers, 248, 280
Comptes rendus, 49–50, 86, 94–95, 117, 198, 243
Compton, Arthur Holly, 223, 239, 280
copper, 85–86
Cotelle, Sonia Slobodkine, 179, 198, 221–22, 238, *266*, 269
Cotton, Aimé, 163
Cotton, Eugénie Feytis, 170, 220–21. (*See also* Feytis Eugénie)
Crompton, Rookes, 33
Crookes, William, 70
curie (radioactivity unit), 109, 232–33
Curie, Eugène (Pierre's father)
  birth and death of, 104, 263, 272
  living with Marie, 49, 60, 69, 80, 97, 98, 101, 121

Richards, Theodore W., 147, 175
Roentgen, Wilhelm, 35, 54–55, 70
Róna, Erzsébet, 218, 271
Rosenblum, Salomon, 198
Rutherford, Ernest
 atomic number-atomic weight connection, 173
 background, 50, 57, 72–73, 76,
  77–78, 141, 182
 birth and death of, 271
 correspondence with Boltwood, 90,
  109, 126, 127–28, 132, 142–43, 191
 correspondence with Gleditsch, 175
 experiments on atomic structure, 138
 lecturing style, 147
 Leslie's work with, 117, 132
 on Marie's honorary degree, 141, 142
 Marie's praise for, 123
 married life, 73, 167, 183
 Nobel Prize winner, 95
 prediction of neutrons, 243–44
 on radium's dangers, 186
 on radium standard, 108–9, 126,
  127–28, 134
 research on emanation (radon),
  73–75, 279
 research on radium, 91, 101, 146
 research on transmutation, 57, 74
 research on uranium, 50, 70, 91
 at Solvay Councils, 119, 120, 253
 on women in research, 183
 World War I and, 165, 167
Rutherford, Mary Newton, 73, 167, 183

Sabin, Florence R., 188, 278
Schrödinger, Erwin, 223
Schützenberger, Paul, 46
séances, 102, 155
Seignobos, Charles, 235–36
Sels de Radium (Radium Salts), 58,
 77, 81, 87, 91, 164, 167. (See also
 Armet de Lisle, factory of)
Sharp, Evelyn, 162

Sikorska, Jadwiga, 145
silicon, 254, 255–56
Sklodowska, Bronya (Marie's sister),
 4–5, 6, 8–9, 10, 11, 14, 272. (See
 also Dluska, Bronya Sklodowska)
Sklodowska, Helena (Marie's sister), 5,
 6, 7, 25, 96, 109, 129, 201, 272
Sklodowski, Józef (Marie's brother)
 birth and death of, 272
 correspondence with Marie on
  family life, 11, 17, 18, 30, 60, 224
 correspondence with Marie on
  health issues, 54, 55
 correspondence with Marie on
  marriage, 24–25, 28
 correspondence with Marie on
  Nobel Prize, 55, 56
 correspondence with Marie on
  Poland, 173
 at Curie Foundation gala, 201
 education, 6
 at Marie's funeral, 258
 at Pierre's funeral, 64
Sklodowska, Marya Salomea
 childhood and early education, 3–6,
  10
 courtship and wedding, 19–26 (See
  also Curie, Marie Sklodowska)
 early employment, 8–10
 magnetism research, 19–21
 as "Marie," 17
 Parisian education, 14–15, 16–19
 vacation with extended family, 6–8, 7
Sklodowski, Wladislaw (Marie's
  father)
 birth and death of, 53, 272
 correspondence with Marie, 31–32,
  53
 devotion to Poland, 3–4
 employment, 3, 4–5, 8
 family ties, 6, 18, 30, 44, 49
 on Marie's love of science, 14
 at Marie's wedding, 25